# Drilling Technology

*Edited by Majid Tolouei-Rad*

Published in London, United Kingdom

IntechOpen

*Supporting open minds since 2005*

Drilling Technology
http://dx.doi.org/10.5772/intechopen.91561
Edited by Majid Tolouei-Rad

Contributors
Hussein Hoteit, Rami Albattat, Edgar Mario Rico Mesa, Carlos Eduardo Palacio Laverde, Carlos Alberto Vergara Crismatt, Sebastian Correa Zapata, Jhon Edison Goez Mora, Juan Camilo Londoño Lopera, Laura Cecilia Tobon Ospina, Yomin Estiven Jaramillo Munera, Juan David Arismendy Pulgarin, John Sneyder Tamayo Zapata, Majid Tolouei-Rad, Muhammad Aamir, Jesus M. Orona-Hinojos

Notice
Statements and opinions expressed in the chapters are these of the individual contributors and not necessarily those of the editors or publisher. No responsibility is accepted for the accuracy of information contained in the published chapters. The publisher assumes no responsibility for any damage or injury to persons or property arising out of the use of any materials, instructions, methods or ideas contained in the book.

First published in London, United Kingdom, 2021 by IntechOpen
IntechOpen is the global imprint of INTECHOPEN LIMITED, registered in England and Wales, registration number: 11086078, 5 Princes Gate Court, London, SW7 2QJ, United Kingdom
Printed in Croatia

British Library Cataloguing-in-Publication Data
A catalogue record for this book is available from the British Library

Additional hard and PDF copies can be obtained from orders@intechopen.com

Drilling Technology
Edited by Majid Tolouei-Rad
p. cm.
Print ISBN 978-1-83968-711-2
Online ISBN 978-1-83968-712-9
eBook (PDF) ISBN 978-1-83968-713-6

We are IntechOpen,
the world's leading publisher of
Open Access books
Built by scientists, for scientists

## 5,300+
Open access books available

## 132,000+
International authors and editors

## 156M+
Downloads

Our authors are among the

## 156
Countries delivered to

## Top 1%
most cited scientists

## 12.2%
Contributors from top 500 universities

Interested in publishing with us?
Contact book.department@intechopen.com

Numbers displayed above are based on latest data collected.
For more information visit www.intechopen.com

# Meet the editor

Dr. Majid Tolouei-Rad is a scientist specializing in manufacturing technologies. He is currently associated with the School of Engineering, Edith Cowan University (ECU), Perth, Western Australia. Since receiving a Ph.D. in 1997, Dr. Tolouei-Rad has been continuously involved in tertiary teaching, research, and supervision of students at undergraduate and postgraduate levels. He has worked for reputable universities in the Middle East, North America, and Oceania, during which he educated dozens of engineers and supervised many engineering projects focusing on engineering materials, manufacturing processes and systems, and automation. He has authored and co-authored more than 80 research articles, books, and book chapters. He has also served on scientific and conference organizing committees and acted as a referee for reputable international journals.

# Contents

# Preface

The range of industries utilizing drilling is wide and includes machining and manufacturing, oil and gas, construction, mining, and so on. The success of drilling operations depends on many factors and much progress has been made in this regard. This book presents the latest scientific and engineering achievements in drilling. It covers new and creative methods of performing this operation, applications of drilling, and process optimization methods to enhance the efficiency and productivity of drilling while lowering associated costs and time. I hope researchers, engineers, and scientists reading this book will find it useful in enhancing the quality of their work and research.

The research presented in this book is that of scientists and engineers from various countries around the world. I am thankful to all the contributors for their work. I would also like to thank Mr. Muhammad Aamir who assisted me in the review process, and Ms. Sara Gojević-Zrnić and the staff at IntechOpen for their continuous support in seeing this book through to publication. Lastly, I give special thanks to my wife, Mahtab Moradi, for her encouragement and support while I was editing this book.

**Dr. Majid Tolouei-Rad**
School of Engineering,
Edith Cowan University,
Perth, Australia

# Chapter 1

# Introductory Chapter: Drilling Technology

*Majid Tolouei-Rad and Muhammad Aamir*

## 1. Introduction

Drilling technology has been widely used in many industries such as manufacturing, mining, oil and gas, and construction. In manufacturing industries, drilling processes are not limited to conventional methods where a physical contact is made between the cutting tool and the solid material. Non-conventional drilling processes use forms of energy such as electrical, chemical, electrochemical, thermal and heat, to generate holes on the hard materials.

Among all drilling processes, conventional drilling with twist drill bits is the first operation attracting extensive attention of researchers. Therefore, the fact that majority of today's products incorporate holes generated by drilling operations cannot be ignored. In various industries, drilling is one of the essential operations, where the joint's life can be critically affected by the quality of the drilled holes. Drilling is often considered the final machining operation during assembly of components, where an efficient drilling process provides superior quality of drilled holes to ensure high strength and high efficiency [1]. A low-quality drilled hole can result in cracks within the structure, which ultimately reduce their service lifetime and add extra costs for maintenance [2]. This is why the drilling process is acknowledged as a more challenging issue during assembly and is the most common, frequent and necessary processing craft in various industries. Therefore, both academia and industries are highly motivated for research on the applications of drilling operations.

However, the drilling process required the use of right cutting tools together with appropriate cutting parameters, such as cutting speed or spindle speed, feed rate, and a reliable machine tool setup, to ensure high quality holes in terms of low surface roughness, cylindricity, circularity, perpendicularity and less formation of burrs [3]. Therefore, the drilling process can be explicitly understood by proper selection and optimization of process parameters without compromising productivity and hole quality [4, 5].

Drilling operations form the largest portion of machining operations in manufacturing industries; therefore, conventional and CNC machines do not give a high production rate where production volumes are huge. In contrast, special purpose machines can provide very high production rates for performing drilling and drilling-related operations including tapping and reaming. The production rates of these machines are many times higher than conventional and CNC machines. In addition, the quality and uniformity of production are superior compared to conventional methods. **Figure 1** shows a two-station special-purpose machine used to perform drilling and tapping operations. The production rate increases as the number of workstations increases, since the machine can perform multiple operations simultaneously. Special purpose machines with 8, 10, and 12 workstations are very common.

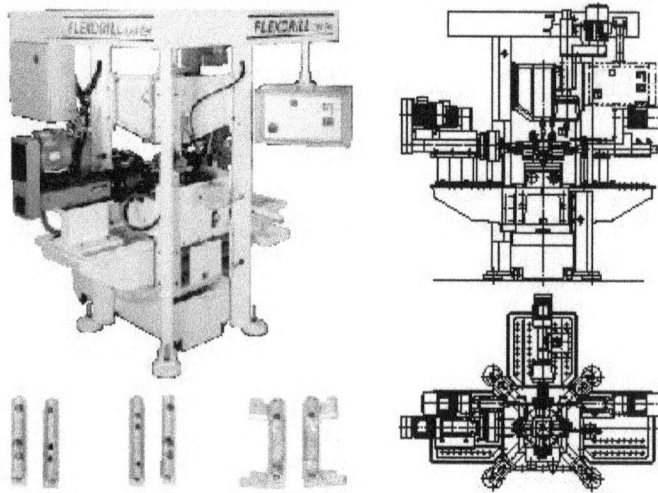

**Figure 1.**
*Flex drill: A special purpose machine used for performing drilling and tapping operations [6].*

**Figure 2.**
*A poly-drill head with three adjustable spindles mounted on conventional milling machine.*

A significant 59% cost reduction is reported in the literature when a special-purpose machine is used in place of a CNC machine, and the cost reduction is an impressive 95.5% when compared to a conventional machine [6]. Although these machines are capable of improving the quality and quantity of the parts produced compared to conventional machines, the utilization of this technology is not proportional to its benefits [6]. This, to a large extent, is attributed to the lack of a solid foundation for feasibility analysis of utilization of these machines. To tackle this, extensive research has been performed by contemporary researchers and models are developed for feasibility analysis of the utilization of such machines; both technically and economically [7–9]. The models developed can assist engineers and manufacturing firms in deciding when this technology gives superior results compared to other alternatives.

Another essential accessory for the drilling operation is the poly-drill head, which can increase production by creating many holes simultaneously [10]. Poly-drill heads are used for drilling and drilling-related operations. As these can perform multiple operations simultaneously, then the overall machining time is reduced significantly, resulting in a huge improved productivity [11, 12]. Poly-drill heads are of fixed and adjustable types, and the number of spindles could vary from two to more than 10. In the adjustable type, the position of the spindles can change providing varied center-to-center distances of the holes within a range. Poly-drills could be used on the conventional drill presses, milling machines, CNC machines, and special-purpose machines. **Figure 2** shows a three-spindle poly-drill head mounted on a conventional milling machine for drilling three holes simultaneously.

Poly drill head also ensures advantages like less rejection of parts by providing better accuracy, less operator fatigue and time saved in the operation. A poly drill head or multi-spindle drill head performed better than the one-shot drilling process by giving a better hole quality, less tool wear, and producing small and fragmented chips [13].

## Conflict of interest

The authors declare no conflict of interest.

Author details

Majid Tolouei-Rad* and Muhammad Aamir
School of Engineering, Edith Cowan University, Joondalup, WA, Australia

*Address all correspondence to: m.rad@ecu.edu.au

IntechOpen

# References

[1] Aamir M, Giasin K, Tolouei-Rad M, Vafadar A. A review: drilling performance and hole quality of aluminium alloys for aerospace applications. Journal of Materials Research and Technology. 2020;9(6):12484-500.

[2] Aamir M, Tolouei-Rad M, Giasin K, Nosrati A. Recent advances in drilling of carbon fiber–reinforced polymers for aerospace applications: a review. The International Journal of Advanced Manufacturing Technology. 2019; 105(5-6):2289-308.

[3] Aamir M, Tolouei-Rad M, Giasin K, Vafadar A. Feasibility of tool configuration and the effect of tool material, and tool geometry in multi-hole simultaneous drilling of Al2024. The International Journal of Advanced Manufacturing Technology. 2020;111(3):861-79.

[4] Karataş MA, Gökkaya H. A review on machinability of carbon fiber reinforced polymer (CFRP) and glass fiber reinforced polymer (GFRP) composite materials. Defence Technology. 2018;In Press.

[5] Aamir M, Tu S, Tolouei-Rad M, Giasin K, Vafadar A. Optimization and modeling of process parameters in multi-hole simultaneous drilling using taguchi method and fuzzy logic approach. Materials. 2020;13(3):680.

[6] Tolouei-Rad M. Intelligent analysis of utilization of special purpose machines for drilling operations. Intelligent Systems, Prof Vladimir M Koleshko (Ed), ISBN: 978-953-51-0054-6, InTech, Available from: http://wwwintechopen com/books/intelligent-systems/ intelligent-analysis-of-utilization-of-special-purposemachines-for-drilling-operations. Croatia2012. p. 297-320.

[7] Vafadar A, Tolouei-Rad M, Hayward K. An integrated model to use drilling modular machine tools. The International Journal of Advanced Manufacturing Technology. 2019; 102(5):2387-97.

[8] Vafadar A, Tolouei-Rad M, Hayward K, Abhary K. Technical feasibility analysis of utilizing special purpose machine tools. Journal of Manufacturing Systems. 2016;39:53-62.

[9] Vafadar A, Hayward K, Tolouei-Rad M. Drilling reconfigurable machine tool selection and process parameters optimization as a function of product demand. Journal of Manufacturing Systems. 2017;45:58-69.

[10] Tolouei-Rad M, Aamir M. Analysis of the Performance of Drilling Operations for Improving Productivity. In: Tolouei-Rad M, editor. Drilling. London, UK: IntechOpen, Available from: https://www.intechopen.com/ online-first/analysis-of-the-performance-of-drilling-operations-for-improving-productivity; 2021.

[11] Aamir M, Tolouei-Rad M, Giasin K, Vafadar A. Machinability of Al2024, Al6061, and Al5083 alloys using multi-hole simultaneous drilling approach. Journal of Materials Research and Technology. 2020;9(5):10991-1002.

[12] Aamir M, Tolouei-Rad M, Vafadar A, Raja MNA, Giasin K. Performance Analysis of Multi-Spindle Drilling of Al2024 with TiN and TiCN Coated Drills Using Experimental and Artificial Neural Networks Technique. Applied Sciences. 2020;10(23):8633.

[13] Aamir M, Tu S, Giasin K, Tolouei-Rad M. Multi-hole simultaneous drilling of aluminium alloy: A preliminary study and evaluation against one-shot drilling process. Journal of Materials Research and Technology. 2020;9(3):3994-4006.

# Analysis of the Performance of Drilling Operations for Improving Productivity

*Majid Tolouei-Rad and Muhammad Aamir*

## Abstract

Drilling is a vital machining process for many industries. Automotive and aerospace industries are among those industries which produce millions of holes where productivity, quality, and precision of drilled holes plays a vital role in their success. Therefore, a proper selection of machine tools and equipment, cutting tools and parameters is detrimental in achieving the required dimensional accuracy and surface roughness. This subsequently helps industries achieving success and improving the service life of their products. This chapter provides an introduction to the drilling process in manufacturing industries which helps improve the quality and productivity of drilling operations on metallic materials. It explains the advantages of using multi-spindle heads to improve the productivity and quality of drilled holes. An analysis of the holes produced by a multi-spindle head on aluminum alloys Al2024, Al6061, and Al5083 is presented in comparison to traditional single shot drilling. Also the effects of using uncoated carbide and high speed steel tools for producing high-quality holes in the formation of built-up edges and burrs are investigated and discussed.

**Keywords:** drilling, cutting tools, hole quality, productivity, multi-spindle head

## 1. Introduction

Drilling is the most commonly performed machining operation in manufacturing industries. Therefore, the analysis and improvement of this process are of great importance in increasing productivity and competitiveness, where many existing studies reported on the optimization and improvement of this process [1, 2]. There are many machines that perform drilling operations including dedicated drilling machines, lathes, milling machines, machining centers and special purpose machines. The drilling process is extensively and heavily used in industries, accounting for a large portion of overall machining time and costs. Therefore, drilling has a significant economic role in industries, where it hugely contributes to the fabrication of various industrial parts [3].

Hole-making processes using drilling operations have been the focus of many research studies, where a lot of development and progress has been made. However, as technology has progressed and newer tools and equipment have been introduced, further research is required to improve the productivity and efficiency of this important operation, which forms the core of activities in many industries [4, 5].

For example, the heat exchangers of nuclear energy centrals require up to 16,000 holes in a single exchanger for assembly with refrigeration tubes [6]. Other examples include the automotive industry, where the drilling process forms up to 40% of total material removed [7], or the aerospace industry where millions of holes are required for joining various parts of aircraft fuselage [8]. It is estimated that 750,000 holes are required in a single wing of an Airbus A380, with 1.5–3 million holes as the requirements for producing a typical commercial aircraft [7]. Furthermore, over a million rivets are needed for large ships [9] where drilling is the primary process. Therefore, a proper selection of machine tools and equipment, cutting tools and parameters is essential in achieving required productivity, dimensional accuracy, and surface roughness. This subsequently helps industries achieve success and improve the service life of their products.

## 2. The drilling process

In the drilling process, holes are created when a cylindrical tool rotates against a workpiece, where a tool called a drill bit is used as shown in **Figure 1** [10, 11]. The drilling operation process involves three stages: the start and centering stage, the full drilling stage and the breakthrough stage [12]. In the first stage, the exact position of the hole is required, whereas the second stage leads to the full engagement of the drill bit, whilst the last stage includes passing the drill through the underside of the workpiece, where the operation stops [13].

A hole in the drilling process can be created in many forms, including blind and through holes, as shown in **Figure 2**. Blind holes are drilled to a certain depth whilst through-holes refer to the condition when the drill bit passes through the material and exits the workpiece on the other side [13].

Generally, a depth to diameter ratio of 5:1 or greater is commonly performed by twist drills, where this ratio may be doubled when using high-performance twist drills equipped with through-tool coolant systems. This ratio can be increased to roughly 20:1 when using special deep hole drilling tools equipped with through-tool coolant systems. Whilst this chapter focuses on the use of twist drills, it is worthnoting that a depth to diameter ratio of 100:1 or more is achievable in gun-drilling machines with through-tool coolant systems. Unlike conventional drilling operations, within gun-drilling machines both the cutting tool and workpiece rotate in opposite directions and at different rotational speeds, which significantly improves the straightness of the deep hole that is generated [14].

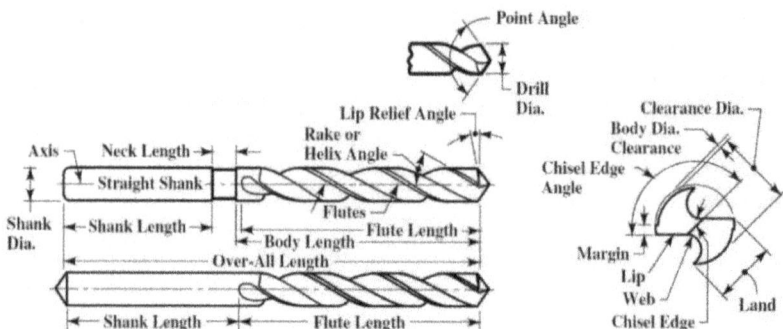

**Figure 1.**
*Standard twist drill nomenclature [11].*

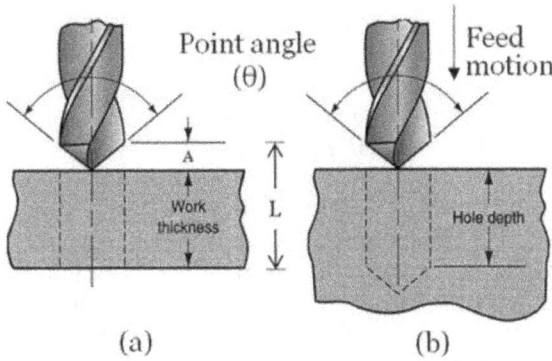

**Figure 2.**
*Drilling process: (a) blind holes (b) through holes [10].*

## 3. Cutting conditions in drilling process

To a large extent, cutting conditions determine the success of any drilling operation. Basic cutting conditions include cutting speed, feed rate, material removal
rate, and machining time, as discussed in this section below.

### 3.1 Spindle speed and cutting speed

The spindle speed is the rotational speed measured in rev/min, calculated using
a tachometer during the drilling process. The spindle speed is used to compute
desired cutting speed, defined as the distance travelled by each cutting edge on the
surface of the workpiece when cutting material. Therefore, cutting speed in a drilling operation is computed by.

$$v = \frac{\pi dn}{1000} \tag{1}$$

where $v$ is the cutting speed in m/min, $\pi = 3.14$, $d$ is the diameter of the
cutting tool in mm, and $n$ is the spindle speed in rev/min.

### 3.2 Feed and feed rate

In a drilling process feed is specified in mm/rev. The feed rate, which is the linear
travel rate in mm/min, can be adjusted by a convenient system when the feed is
multiplied by the spindle speed. Hence, feed rate can be found as.

$$f_r = fn \tag{2}$$

where $f_r$ is the feed rate in mm/min, $f$ is the feed in mm/rev, and $n$ is the
spindle speed in rev/min.

### 3.3 Material removal rate

The material removal rate can be considered as an index for the determination
of the efficiency of a machining process. In a drilling process, material removal is
obtained by [15].

$$M_{rr} = \left(\frac{\pi}{4}\right) d^2 f_r \tag{3}$$

where $M_{rr}$ is the material removal rate in mm³, $d$ is the diameter of the drill in mm, and $f_r$ is the feed rate in mm/min.

### 3.4 Drilling time

Drilling time is the time a tool is engaged from the beginning of chip production to the end for uninterrupted machining. Any pause during this process, either planned or unplanned, is not included in this time. The drilling time in minutes for through holes can be determined by [15].

$$T_m = \frac{L}{f_r} \tag{4}$$

where $T_m$ is drilling time in minutes, $L$ is the distance travelled by the cutting tool in mm, and $f_r$ is the feed rate in mm/min.

It should be noted that the drill bit should travel the distance $L$ (see **Figure 2**), which consists of the desired depth of the hole plus an allowance for the tool point angle, $A$, given by.

$$A = \frac{d}{2} \tan\left(90 - \frac{\theta}{2}\right) \tag{5}$$

where $A$ is the allowance in mm, $d$ is the diameter of the drill in mm, and $\theta$ is the tool point angle in degrees.

## 4. Aluminium alloys

Aluminium and its alloys are very attractive to many manufacturing industries due to its unique combinations of properties with outstanding engineering applications across various industries [16, 17]. Aluminium has low density, reasonably high strength, high ductility, high thermal and electrical conductivities, good oxidation and corrosion resistance, easy to manufacture and has a relatively low cost [18].

The high strength-to-weight ratio of aluminium alloys makes them suitable for wide use in marine, automotive and aerospace industries [19]. The various grades of aluminium alloys used in the aviation industry can be found in reference [5]. Aluminium and its alloys are also used in home appliances, construction industries, electrical, electronic, packaging industries, etc. [16, 19].

Aluminium alloys are divided into workable alloys and cast alloys. Alloys of aluminium that undergo hot or cold mechanical working processes are termed as workable alloys, while those whose shape is obtained by the casting process are known as cast alloys [19].

Aluminium alloys are generally considered more machinable than ferrous alloys; however, their ductile nature results in high machining forces, poor surface roughness and difficult control of chips, whereas those with hard particles can cause high tool wear [19].

## 5. Multi-spindle drilling for productivity improvement

Multi-spindle drilling is used in manufacturing industries to improve productivity as they can drill many holes simultaneously, which reduces machining time significantly. The multi-spindle or poly-drill head gives high center-to-center accuracy and in many instances eliminates the need for the use of drilling jigs, and this further decreases drilling time and cost. Therefore, in today's competitive market, it is essential to produce a large number of products at the right time with high quality and at minimum cost, where the use of a multi-spindle drill head is one way to fulfil this goal. A multi-spindle drill head can simultaneously drill from two to ten or more holes on the same plane [20]. The multi-spindle drill head produces a number of holes of similar quality in the most economical way, providing a high level of automation with a small investment [21].

Multi-spindle drilling technology is used to increase productivity whilst reducing machining time in working conditions where a large number of closely-spaced holes need to be drilled. A good example of this is the manufacturing of aircraft fuselage and construction of metal bridges, where a large number of riveting holes are required. **Figure 3** shows a section of the Golden Gate Bridge which has been constructed using a large number of rivets. It is estimated that approximately 600,000 rivets are used in this structure [22].

Multi-spindle or poly-drill heads are mounted on a machine tool to perform many operations simultaneously [23]. Multi-spindle drill heads are either fixed or flexible. The tool positions in fixed multi-spindle drill heads cannot change. Whereas in the flexible type the positions of tools can be adjusted as needed within a particular range [24]. **Figure 4** shows an adjustable 3-spindle drill head [25]. The importance of using this poly drill head instead of using a single drill bit is the possibility of producing high quality drilled holes, the elimination of a drilling jig for maintaining a high center-to-center tolerance, fewer rejections, reducing

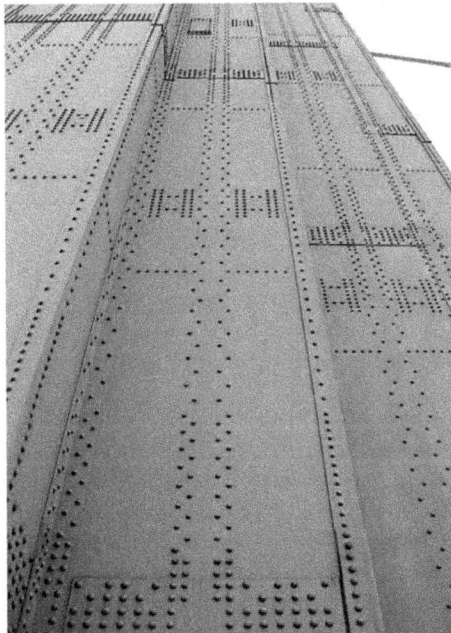

**Figure 3.**
*Thousands of rivets are used in the structure of the Golden Gate Bridge, San Francisco, United States.*

**Figure 4.**
*An adjustable 3-spindle drill head [25].*

machining time, increasing profit rate and less operator fatigue [26]. Therefore, it is worth noting that the use of multi-spindle drilling is an excellent choice to improve productivity and reduce machining time for manufacturing industries requiring the production of a large number of holes with stringent tolerances. Therefore, the advantages of using the multi-spindle head are listed below [23]:

- The increase in productivity at a higher rate

- The performance of multiple operations in one cycle

- The time for one hole is the time for multiple numbers of holes

- The multi-spindle drilling ensures positional accuracy

- Elimination or reduction of the need for drilling jigs

- Less quality control rejections

- Easy to install and use anywhere

- Easy to operate and low maintenance

- Simple in construction and robust in design

### 5.1 Cutting mechanisms in the drilling of aluminium

In the machining process, when a tool penetrates inside a metal workpiece, it produces an internal shearing action in the metal where the metal becomes severely stressed. This causes the metal to be plastically deformed and flow in the form of chips when the ultimate shear strength of the metal is exceeded [27]. In the drilling process, the thrust force is the perpendicular force to the workpiece during its translational motion while the torque comes from the machine spindle to rotate the tool during drilling operation. Other forces in drilling are not important as they are small compared to the thrust force [28]. It should be noted that high cutting forces affect hole quality and tool life [29]. Forces generated in the drilling of metals are uniform where uncut chip thickness is constant [30].

Experimental studies have shown that thrust force generated in multi-spindle drilling is higher than that obtained in one-shot single drilling processes of aluminum alloy Al5083 [31]. The higher thrust force occurs due to the combination of more than one tool operating simultaneously in one go. However, the results of experiments have concluded that the average of all the tools' thrust force per tool in multi-spindle drilling was slightly lower than the thrust force resulting from single drilling [31]. In addition to thrust force, another important parameter in a drilling process is the increase in cutting temperature [32]. A higher cutting temperature increases the ductility of the material which results in the formation of long chips, which negatively affects the hole quality [33]. A high temperature may also increase the chemical interaction between aluminium and the tool coating that is responsible for inter-atomic diffusion [34]. The cutting temperature increases due to heat generation which is the result of an increase in cutting speed [35].

Further, in machining ductile materials like aluminium, there is a chance of producing continuous chips due to the plastic deformation of its ductile nature. Other factors that contribute to the formation of continuous chips are high cutting speed, sharp cutting edge, etc. Continuous chips are not easy to handle and dispose of, where they can get tangled around the tool and pose safety issues to the operator. Additionally, when a tool face is in contact for a long time, it results in more frictional heat and affects machining. Therefore, discontinuous and segmented chips produce less friction between the tool and chip; hence, resulting in a better surface finish and providing higher operator safety [27].

## 5.2 Cutting tool and spindle adjustment in multi-spindle drilling of aluminium alloys

As mentioned earlier, the best performance in a drilling process is obtained when using appropriate cutting tools, where the correct process conditions are used to reduce the level of damage as much as possible [27]. The most commonly used drill bit is the twist drill, as shown in **Figure 1**, which represents the industrial standard [36]. The important features of the twist drill include the point angle, clearance angle, chisel edge angle, drill diameter, web thickness, the rake angle, etc. The rake angle in drills is specified as the helix angle [13]. The high point angle and large helix angle are recommended for better hole quality and less tool wear [37]. The large point angle also contributes to producing thinner chips during the machining of aluminium alloys [38]. However, the point angle should be selected based on silicon contents in aluminium alloys [39].

Experimental studies performed in references [31, 40] have shown that tools used for multi-spindle drilling give less formation of built-up edges as compared to the single drilling process of aluminum alloy Al5083 due to differences in chip size when uncoated High-Speed Steel (HSS) drills with a point angle of 118° and size of 6 mm were used. The experiments were conducted using a conventional milling machine for both single drilling and multi-spindle drilling processes, and the same drilling parameters and conditions were applied. For the multi-spindle drilling process, a SUNHER poly-drill head, as shown in **Figure 5**, was used.

Multi-spindle drilling experiments were further extended and uncoated HSS drills were tested on aluminium alloy Al2024 and compared with uncoated carbide drills with a point angle of 140° and a diameter of 6 mm. Apart from aluminum alloy Al2024, the 6 mm uncoated carbide drills were also used for multi-spindle drilling of aluminium alloys Al5083 and Al6061. In addition, 6 mm uncoated carbide drills were used to compare different center-to-center tool distances of the spindle in the multi-spindle drilling process. Further, a comparison of 6 mm and 10 mm uncoated carbide drills with the same point angle of 140° were also made [41, 42].

**Figure 5.**
*The 3-spindle Suhner multi-spindle drill head mounted on conventional milling machine (Courtesy: Edith Cowan University, Australia).*

In general, the uncoated carbide drill has been recommended in multi-spindle drilling of aluminium as compared to the uncoated HSS drills due to the high built-up edge formation because of its moderate strength, as shown in **Figure 6**. The drill diameter did not show any significant changes in affecting the hole quality; however, the larger drill size covered a larger cross-sectional area that resulted in a higher thrust force and producing larger chips. Therefore, for the smaller drill size, an easier chip breaking and evacuation was resulted. Furthermore, the larger point angle of 140° - compared to 118° - provided a better hole quality but did not contribute to changing the size or shape of the chips.

The tool conditions from **Figure 6** also shows that when drilling aluminium alloy Al5083, a large built-up edge was formed, which was expected due to low silicon contents, where this is in agreement with research conducted by Akyüz [43] in which alloys with low silicon contents produce a high built-up edge. Additionally, the low hardness value of the material used in this operation might be another cause of high formation of the built-up edge because alloys with low hardness values have a high tendency towards the formation of built-up edges [44].

Multi-spindle drilling is useful in its easy adjustment of tools. Depending on the type and use, the tools of a multi-spindle head can be adjusted to any position without affecting the results, which not only increases productivity at a high rate but also produces high-quality holes. This is performed at the same time, whereas only a single hole is produced in one-shot single drilling process without a compromise on the hole quality.

### 5.3 Quality assessment of drilled holes in multi-spindle drilling of aluminium

In any drilling process, it is important to ensure that damage-free and precise holes are produced to avoid rejection of parts [45]. For example, poor hole quality has been observed in 60% of aircraft components [5], which is of course a

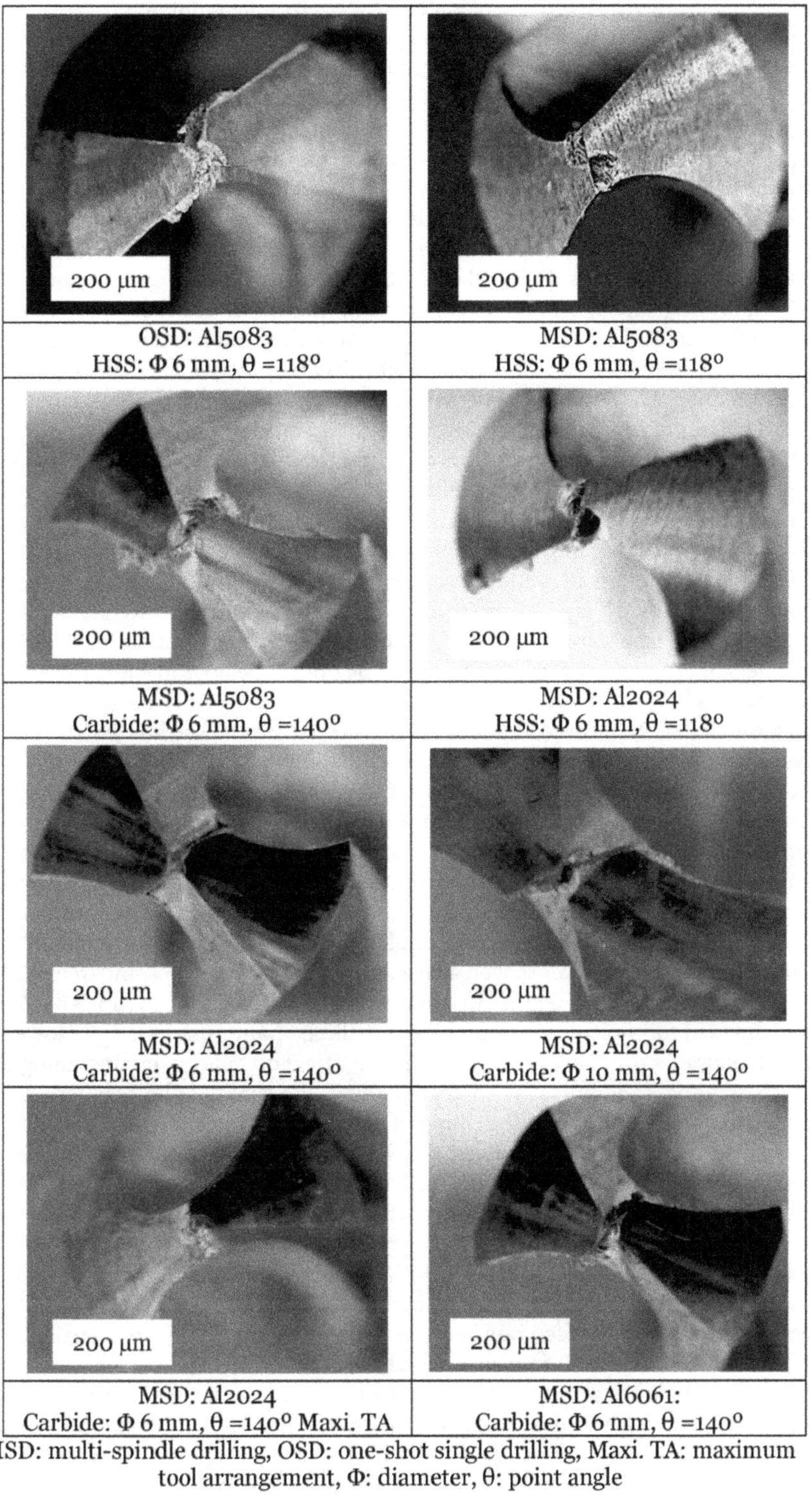
200 μm
OSD: Al5083
HSS: Φ 6 mm, θ =118°
200 μm
MSD: Al5083
HSS: Φ 6 mm, θ =118°
200 μm
MSD: Al5083
Carbide: Φ 6 mm, θ =140°
200 μm
MSD: Al2024
HSS: Φ 6 mm, θ =118°
200 μm
MSD: Al2024
Carbide: Φ 6 mm, θ =140°
200 μm
MSD: Al2024
Carbide: Φ 10 mm, θ =140°
200 μm
MSD: Al2024
Carbide: Φ 6 mm, θ =140° Maxi. TA
200 μm
MSD: Al6061:
Carbide: Φ 6 mm, θ =140°
MSD: multi-spindle drilling, OSD: one-shot single drilling, Maxi. TA: maximum tool arrangement, Φ: diameter, θ: point angle

**Figure 6.**
*Tool conditions after performing the drilling operation on different types of aluminium alloys using the multi-spindle drill head.*

challenging problem. Hence, there is a need to control the number of rejected parts by overcoming problems related to the drilling process, especially the quality of drilled-holes [45]. A poor quality hole can create regions of concentrated stress that increase the chances of formation of fatigue cracks, which reduces the reliability of products [46]. Desirable hole quality in drilling operations can be achieved by proper selection of drilling process parameters, appropriate cutting tools, and machine setup [47].

In the experimental study by Aamir et al. [31], a single drilling process was compared with multi-spindle simultaneous drilling of aluminium alloy Al5083 using uncoated HSS tools. The drill diameter was 6 mm and the point angle was 118°. All drilling experiments were conducted using the same cutting parameters and in a dry environment. The hole quality produced by multi-spindle drilling was better than those obtained in a single-spindle drilling process. The holes drilled by the multi-spindle head had a lower surface roughness and fewer burrs around the holes. This was expected due to differences in chip formation.

Hole quality in a drilling process is also affected by the chemical composition and mechanical properties of aluminium alloys. An experimental study in multi-spindle drilling of aluminium alloys Al5083, Al6061, and Al2024 by Aamir et al. [42] concluded that regardless of drilling parameters, low surface roughness was obtained in aluminium alloy Al6061 due to its high silicon content. Literature has shown that alloys with high silicon content result in low surface roughness irrespective of drilling parameters [43, 48]. Furthermore, the reason for high surface roughness of aluminium alloy Al5083 might be due to the poor machinability and low hardness [44].

Aamir et al. [42] observed that less burrs formed around the hole edges of aluminium alloy Al2024 due to its good machinability compared with the aluminium alloys Al6061 and Al5083. Further, the less ductile nature of aluminium alloy Al6061 - in comparison with aluminium alloy Al5083 - resulted in the low formation of burrs. Hence, high ductility and poor machinability properties led to the formation of more burrs in aluminium alloy Al5083 [49]. Additionally, uncoated HSS drills produced low-quality holes by giving high surface roughness and more formation of burrs around the edges of holes due to the high built-up edges because of its moderate strength. Moreover, a larger point angle of 140° - compared to 118°- and a smaller diameter of 6 mm - in comparison to a drill size of 10 mm – have been recommended for multi-hole simultaneous drilling of aluminium alloys. However, the drill diameter did not show any significant effect on hole quality including the surface roughness and burrs [41]. **Figure 7** shows the quality of holes in terms of burr formation in multi-spindle drilling of aluminium alloys.

Regardless of drilling parameters, the workpiece materials and different tools, the surface roughness increases with increasing the cutting speed and feed rate. The likely reasons for high surface roughness at high cutting speeds might include the increase in workpiece deformation due to rise in temperature and the chances of high vibrations exerted by the tools [7, 47]. The high cutting speed and feed rate are responsible for the formation of burrs that reduces the hole quality. However, the high impact is due to the feed rate because stable, jerk-free and slow insertions of drills are possible with low feed rates which form thin chips; hence, the hole quality is less affected [50]. Further, according to Costa et al. [51], any factor that causes the generation of high thrust force results in more formation of burrs, and the high feed rate increases thrust force. Additionally, due to less formation of burrs, the tool entry side of the holes was found to be better than those on the tool exit side. This is likely due to the different mechanism of burr formation at the entry and exit sides of the holes. According to Zhu et al. [52], the tearing occurs as a bending action followed by clean shearing or lateral extrusion causing entrance burrs while exit

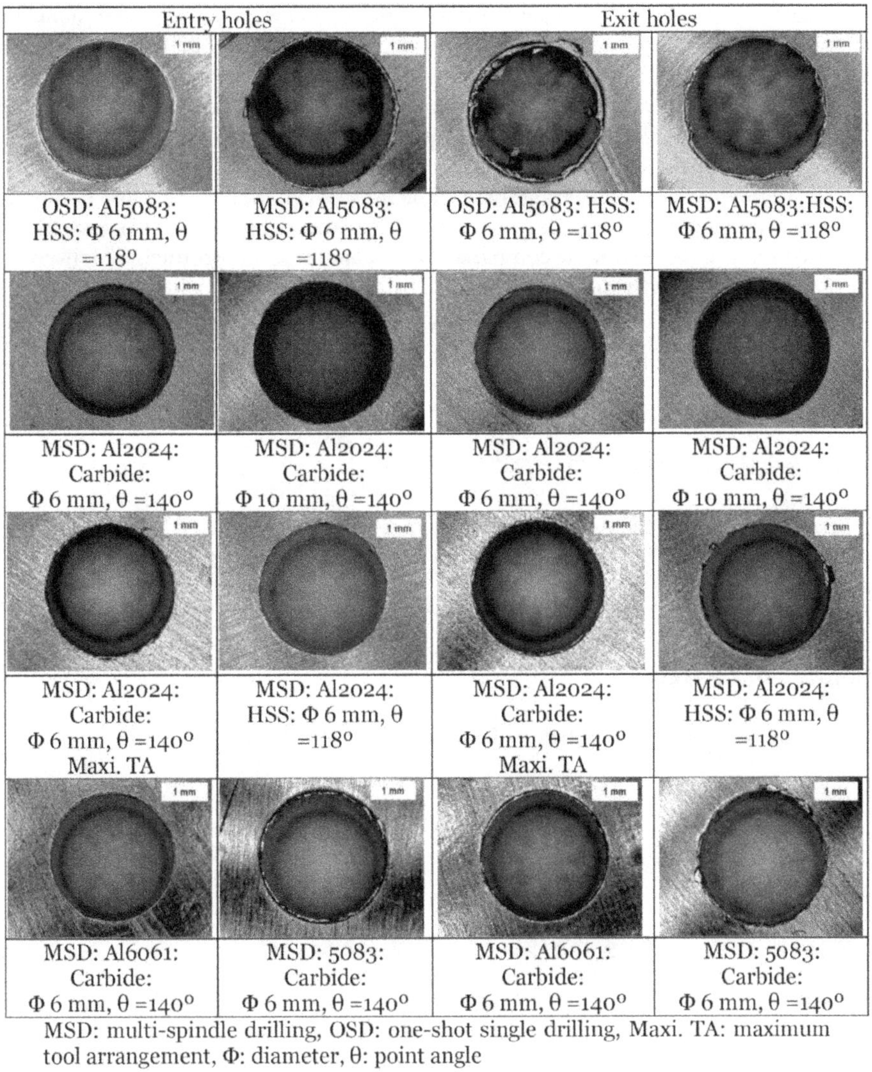

| Entry holes | | Exit holes | |
|---|---|---|---|
| OSD: Al5083: HSS: $\Phi$ 6 mm, $\theta$ =118° | MSD: Al5083: HSS: $\Phi$ 6 mm, $\theta$ =118° | OSD: Al5083: HSS: $\Phi$ 6 mm, $\theta$ =118° | MSD: Al5083:HSS: $\Phi$ 6 mm, $\theta$ =118° |
| MSD: Al2024: Carbide: $\Phi$ 6 mm, $\theta$ =140° | MSD: Al2024: Carbide: $\Phi$ 10 mm, $\theta$ =140° | MSD: Al2024: Carbide: $\Phi$ 6 mm, $\theta$ =140° | MSD: Al2024: Carbide: $\Phi$ 10 mm, $\theta$ =140° |
| MSD: Al2024: Carbide: $\Phi$ 6 mm, $\theta$ =140° Maxi. TA | MSD: Al2024: HSS: $\Phi$ 6 mm, $\theta$ =118° | MSD: Al2024: Carbide: $\Phi$ 6 mm, $\theta$ =140° Maxi. TA | MSD: Al2024: HSS: $\Phi$ 6 mm, $\theta$ =118° |
| MSD: Al6061: Carbide: $\Phi$ 6 mm, $\theta$ =140° | MSD: 5083: Carbide: $\Phi$ 6 mm, $\theta$ =140° | MSD: Al6061: Carbide: $\Phi$ 6 mm, $\theta$ =140° | MSD: 5083: Carbide: $\Phi$ 6 mm, $\theta$ =140° |

MSD: multi-spindle drilling, OSD: one-shot single drilling, Maxi. TA: maximum tool arrangement, $\Phi$: diameter, $\theta$: point angle

**Figure 7.**
*Hole quality in terms of burr formation in multi-spindle drilling of aluminium alloys using the multi-spindle head.*

burrs formed as a result of plastic deformation of the materials. Besides this, low temperature and thrust force are the reasons for small burrs on the entry side of the hole which can be removed by chamfering [53].

## 6. Conclusions

Many industries, such as automotive and aerospace, produce millions of holes per day where productivity, quality, and precision of drilled holes plays a vital role in their success. Multi-spindle drilling is capable of producing more drilled-holes with higher rates, which makes it advantageous in high-volume production with uniform qualities, simultaneous machining, and most importantly reducing the drilling time which is one of the important factors in achieving greater productivity.

Therefore, the production of a large number of closely-spaced holes simultaneously by using a poly-drill or multi-spindle drill head results in achieving higher productivity and quality. This approach not only enhances the competitiveness of the process but also results in cost reduction and uniformity of generated holes.

Further, it can be concluded that hole quality is affected by drilling parameters and properties of the workpiece. Alloys of aluminium with high silicon contents show lower values of surface roughness while those with low hardness and poor machinability provide poor hole quality. The experiments also show that uncoated carbide tools are more suitable compared to uncoated HSS for producing high-quality holes, resulting in the formation of less built-up edges when drilling aluminium alloys. Moreover, the aluminium alloy Al2024 produced better results in terms of hole quality due to its good machinability compared with aluminum alloys Al6061 and Al5083.

## Acknowledgements

The authors would like to thank Edith Cowan University, Australia for the awarded (ECU-HDR) higher degree research scholarship and for providing support on this research.

## Conflict of interest

The authors declare no conflict of interest.

Author details

Majid Tolouei-Rad* and Muhammad Aamir
School of Engineering, Edith Cowan University, Joondalup, WA, Australia

*Address all correspondence to: m.rad@ecu.edu.au

IntechOpen

# References

[1] Vafadar A, Hayward K, Tolouei-Rad M. Drilling reconfigurable machine tool selection and process parameters optimization as a function of product demand. Journal of Manufacturing Systems. 2017;45:58-69.

[2] Tolouei-Rad M, Shah A. Development of a methodology for processing of drilling operations. International Journal of Industrial and Manufacturing Engineering. 2012;6(12):2660-2664.

[3] Kilickap E. Modeling and optimization of burr height in drilling of Al-7075 using Taguchi method and response surface methodology. The International Journal of Advanced Manufacturing Technology. 2010;49(9-12):911-923.

[4] Aamir M, Tolouei-Rad M, Giasin K, Nosrati A. Recent advances in drilling of carbon fiber–reinforced polymers for aerospace applications: a review. The International Journal of Advanced Manufacturing Technology. 2019;105(5-6):2289-2308.

[5] Aamir M, Giasin K, Tolouei-Rad M, Vafadar A. A review: drilling performance and hole quality of aluminium alloys for aerospace applications. Journal of Materials Research and Technology. 2020;9(6):12484-12500.

[6] De Lacalle LL, Fernández A, Olvera D, Lamikiz A, Rodríguez C, Elias A. Monitoring deep twist drilling for a rapid manufacturing of light high-strength parts. Mechanical systems and signal processing. 2011;25(7):2745-2752.

[7] Giasin K, Hodzic A, Phadnis V, Ayvar-Soberanis S. Assessment of cutting forces and hole quality in drilling Al2024 aluminium alloy: experimental and finite element study. The International Journal of Advanced Manufacturing Technology. 2016;87(5-8):2041-2061.

[8] Rivero A, Aramendi G, Herranz S, de Lacalle LL. An experimental investigation of the effect of coatings and cutting parameters on the dry drilling performance of aluminium alloys. The International Journal of Advanced Manufacturing Technology. 2006;28(1-2):1-11.

[9] Felkins K, Leigh H, Jankovic A. The royal mail ship Titanic: Did a metallurgical failure cause a night to remember? JOM Journal of the Minerals, Metals and Materials Society. 1998;50(1):12-18.

[10] Groover M. Fundamental of modern manufacturing: Materials,Processes, andSystems. USA: John Wiley & Sons, Inc; 2004.

[11] Oberg E. Machinery's Handbook 29th Edition-Full Book: Industrial Press; 2012.

[12] Tönshoff H, Spintig W, König W, Neises A. Machining of holes developments in drilling technology. CIRP annals. 1994;43(2):551-561.

[13] Sharif S, Rahim EA, Sasahara H. Machinability of titanium alloys in drilling. Titanium Alloys-Towards Achieving Enhanced Properties for Diversified Applications. 32012. p. 117-137.

[14] Systems UDHD. What is Gun Drilling? 2020. Available from: https://unisig.com/information-and-resources/what-is-deep-hole-drilling/what-is-gun-drilling/.

[15] Girsang IP, Dhupia JS. Machine Tools for Machining. In: Nee AYC, editor. Handbook of Manufacturing Engineering and Technology. London: Springer London; 2015. p. 811-865.

[16] David J. Aluminum and Aluminum alloys. Alloying: Understanding the

basics. ASM International, Ohio; 2001. p. 351-416.

[17] Aamir M, Tolouei-Rad M, Vafadar A, Raja MNA, Giasin K. Performance Analysis of Multi-Spindle Drilling of Al2024 with TiN and TiCN Coated Drills Using Experimental and Artificial Neural Networks Technique. Applied Sciences. 2020;10(23):8633.

[18] Campbell FC. Aluminum. Elements of Metallurgy and Engineering Alloys: ASM International; 2008. p. 487-508.

[19] Santos MC, Machado AR, Sales WF, Barrozo MA, Ezugwu EO. Machining of aluminum alloys: a review. The International Journal of Advanced Manufacturing Technology. 2016;86(9-12):3067-3080.

[20] Tolouei-Rad M. An intelligent approach to high quantity automated machining. Journal of Achievements in Materials and Manufacturing Engineering. 2011;47(2):195-204.

[21] Tolouei-Rad M. Intelligent analysis of utilization of special purpose machines for drilling operations. Intelligent Systems, Prof Vladimir M Koleshko (Ed), ISBN: 978-953-51-0054-6, InTech, Available from: http://wwwintechopencom/books/intelligent-systems/intelligent-analysis-of-utilization-of-special-purposemachines-for-drilling-operations. Croatia2012. p. 297-320.

[22] Golden Gate Bridge HaTD. Frequently Asked Questions about the Golden Gate Bridge: How many rivets are in each tower of the Golden Gate Bridge 2020. Available from: https://www.goldengate.org/bridge/history-research/statistics-data/faqs/.

[23] Tam BN, Van Dich T. Research, Design and Develop a prototype of Multi-Spindle Drilling Head. Journal of Science & Technology. 2018;127:029-034.

[24] Tolouei-Rad M. An efficient algorithm for automatic machining sequence planning in milling operations. International Journal of Production Research. 2003;41(17):4115-4131.

[25] Tolouei-Rad M, Zolfaghari S. Productivity improvement using Special-Purpose Modular machine tools. International Journal of Manufacturing Research. 2009;4(2):219-235.

[26] Vafadar A, Tolouei-Rad M, Hayward K, Abhary K. Technical feasibility analysis of utilizing special purpose machine tools. Journal of Manufacturing Systems. 2016;39:53-62.

[27] Sobri SA, Heinemann R, Whitehead D. Carbon Fibre Reinforced Polymer (CFRP) Composites: Machining Aspects and Opportunities for Manufacturing Industries. Composite Materials: Applications in Engineering, Biomedicine and Food Science. Cham: Springer International Publishing; 2020. p. 35-65.

[28] Tyagi R. Processing Techniques and Tribological Behavior of Composite Materials: IGI Global; 2015.

[29] Xu J, Mkaddem A, El Mansori M. Recent advances in drilling hybrid FRP/Ti composite: a state-of-the-art review. Composite Structures. 2016;135:316-338.

[30] Sheikh-Ahmad JY. Machining of polymer composites: Springer; 2009.

[31] Aamir M, Tu S, Giasin K, Tolouei-Rad M. Multi-hole simultaneous drilling of aluminium alloy: A preliminary study and evaluation against one-shot drilling process. Journal of Materials Research and Technology. 2020;9(3):3994-4006.

[32] Kelly J, Cotterell M. Minimal lubrication machining of aluminium alloys. Journal of Materials Processing Technology. 2002;120(1-3):327-334.

[33] Ozcatalbas Y. Chip and built-up edge formation in the machining of in situ Al4C3–Al composite. Materials & design. 2003;24(3):215-221.

[34] Roy P, Sarangi S, Ghosh A, Chattopadhyay A. Machinability study of pure aluminium and Al–12% Si alloys against uncoated and coated carbide inserts. International Journal of Refractory Metals and Hard Materials. 2009;27(3):535-544.

[35] Yousefi R, Ichida Y. A study on ultra–high-speed cutting of aluminium alloy:: Formation of welded metal on the secondary cutting edge of the tool and its effects on the quality of finished surface. Precision engineering. 2000;24(4):371-376.

[36] Panchagnula KK, Palaniyandi K. Drilling on fiber reinforced polymer/nanopolymer composite laminates: a review. Journal of materials research and technology. 2018;7(2):180-189.

[37] Nouari M, List G, Girot F, Coupard D. Experimental analysis and optimisation of tool wear in dry machining of aluminium alloys. Wear. 2003;255(7-12):1359-1368.

[38] Stephenson DA, Agapiou JS. Metal cutting theory and practice: CRC press; 2005.

[39] Davim JP. Modern machining technology: A practical guide. UK: Elsevier; 2011. 412 p.

[40] Aamir M, Tu S, Tolouei-Rad M, Giasin K, Vafadar A. Optimization and modeling of process parameters in multi-hole simultaneous drilling using taguchi method and fuzzy logic approach. Materials. 2020;13(3):680.

[41] Aamir M, Tolouei-Rad M, Giasin K, Vafadar A. Feasibility of tool configuration and the effect of tool material, and tool geometry in multi-hole simultaneous drilling of

Al2024. The International Journal of Advanced Manufacturing Technology. 2020;111(3):861-879.

[42] Aamir M, Tolouei-Rad M, Giasin K, Vafadar A. Machinability of Al2024, Al6061, and Al5083 alloys using multi-hole simultaneous drilling approach. Journal of Materials Research and Technology. 2020;9(5):10991-11002.

[43] Akyüz B. Effect of silicon content on machinability of AL-SI alloys. Advances in Science and Technology Research Journal. 2016;10(31):51--57.

[44] Ratnam M. Factors affecting surface roughness in finish turning. In: Comprehensive materials finishing. Elsevier. 2017;1(1):1-25.

[45] Arul S, Vijayaraghavan L, Malhotra S, Krishnamurthy R. The effect of vibratory drilling on hole quality in polymeric composites. International Journal of Machine Tools and Manufacture. 2006;46(3-4):252-259.

[46] Liu J, Xu H, Zhai H, Yue Z. Effect of detail design on fatigue performance of fastener hole. Materials & Design. 2010;31(2):976-980.

[47] Kurt M, Kaynak Y, Bagci E. Evaluation of drilled hole quality in Al 2024 alloy. The International Journal of Advanced Manufacturing Technology. 2008;37(11-12):1051-1060.

[48] Kamiya M, Yakou T, Sasaki T, Nagatsuma YJMt. Effect of Si content on turning machinability of Al-Si binary alloy castings. 2008:0801280304-.

[49] Committee AH. Metals Handbook: Vol. 2, Properties and selection–nonferrous alloys and pure metals. American Society for Metals, Metals Park, OH. 1978.

[50] Uddin M, Basak A, Pramanik A, Singh S, Krolczyk GM, Prakash C. Evaluating hole quality in drilling of Al 6061 alloys. Materials. 2018;11(12):2443.

[51] Costa ES, Silva MBd, Machado AR. Burr produced on the drilling process as a function of tool wear and lubricant-coolant conditions. Journal of the Brazilian Society of Mechanical Sciences and Engineering. 2009;31(1):57-63.

[52] Zhu Z, Guo K, Sun J, Li J, Liu Y, Zheng Y, et al. Evaluation of novel tool geometries in dry drilling aluminium 2024-T351/titanium Ti6Al4V stack. Journal of Materials Processing Technology. 2018;259:270-281.

[53] Shanmughasundaram P, Subramanian R. Study of parametric optimization of burr formation in step drilling of eutectic Al–Si alloy–Gr composites. Journal of Materials Research and Technology. 2014;3(2):150-157.

Chapter 3

# Innovative Double Cathode Configuration for Hybrid ECM + EDM Blue Arc Drilling

*Jesus M. Orona-Hinojos*

Abstract

Electrical discharge machining is a machining method generally used for machining hard metals, those that would be high cost or have poor performance to machine with other techniques using, e.g., lathes, drills, or conventional machining. Therefore, also known as thermal processes like EDM, Plasma or Laser cutting can be used in drilling operations with poor metallurgical quality on cutting edge and will be necessary complement with other processes such as electrochemical machining (ECM). Both ECM and EDM processes use electrical current under direct-current (DC) voltage to electrically power the material removal rate (MRR) from the workpiece. However in ECM, an electrically conductive liquid or electrolyte is circulated between the electrode(s) and the workpiece for permitting electrochemical dissolution of the workpiece material. While the EDM process, a nonconductive liquid or dielectric is circulated between the cathode and workpiece to permit electrical discharges in the gap there between for removing the workpiece material. Both are principle too different, EC using an electrical conductive and ED using a dielectric medium. But exist a way that can to do a combination of Pulsed EC + ED Simultaneous and allowing the coexist both process, in a semidielectric medium, where both condition exist in the same time, therefore in this hybrid is possible create a tooling device dual cathode for drilling process with promissory advantages fast hole for this innovative hybrid ECDM Simultaneous, this hybrid it's knew as blue arc drilling technology.

**Keywords:** Drilling, EDM and ECM process, Double Cathode, Hybrid ECDM

## 1. Introduction

Is known that ECM and EDM are machining processes that each one has reached a maximum in the material removal rate (MMR), mainly due to conditions of electrochemical and physical equilibrium respectively. These processes defined as electrical discharge machining (EDM) and electrochemical machining (ECM) have high adaptability to make some variants of assisted hybrid systems that allow the acceleration of mass transport to improve considerably the metal removal rate measured in $mm^3/min$. Then those processes present important advantages when combined with other variants such as the use of abrasive materials (G), ultrasound (US), laser projection (LB), and hydrodynamic magnetic force (HMF) that scientific community has reported in the last decades, then the integration is a possible challenge for engineers and technologists today.

IntechOpen

The evidence on the growth of these removal speeds regain the interest of the industrial sector, being the advanced hybrid machining processes (HMP) like to EDCM technology that allows them will be competitive on some parameters versus Laser or Plasma thermal cutting with high material removal rate, but with a severe heat-affected zone (HAZ), between 1000 to 1600 μm. While in non-contact cutting processes EDM and ECM the HAZ is minimized below 40 μm. However, in terms of material removed, the ECM has speeds of the order of 100 to 250 mm$^3$/min, depending on the work material and current density among other parameters. EDM process, the removal speed is between 300 to 600 mm$^3$/min, depending on the discharge power and duty cycle [1].

Variants to the recently published non-contact machining processes [2, 3] open up new lines of innovation in the use of hybrid high-speed EDM technology in drilling and grinding for: (i) Multi-manufacturing of complex precision 3D with additive-laser. (ii) Manufacture of alloys high strength with friction-free finishes (Ra < 400 nm). An EDM electro-discharge erosion process, also known as Blasting BEAM (Blasting Erosion Arc Machining) [4] is reported, with MRR of the order of 11,000 mm$^3$/min in Inconel 1718, obtained experimentally.

General Electric Inc. in 2011 showed evidence of technological development of machining for high-speed blades with hybrid EDM in low thermal impact named Blue Arc Machining a device with registration US2010/0126877 A1. GE's laboratory in China achieves MRR in the order of 3,500 to 5,000 mm$^3$/min.

On the other hand, leading global companies in thermal cutting processes such as TRUMPF Inc. unveiled in 2011, a hybrid Laser/EDM Drilling Cell, with removal capacities of 30,000 to 35,000 mm$^3$/min, depending on the type of part to be manufactured "light-alloy", "medium-alloy" or "duty-alloy" component patent registration EP1988/0299143 A1 [5]. In other words, conventional laser and plasma thermal processes are also reaching their removal speed limit and are evolving into special hybrids.

A couple of decades ago, advanced materials and cutting precision were intended for components of the aerospace and aeronautical sector, which for safety were manufactured piece by the piece it's known as "aircraft-components", and the manufacturing precision allowed by electro-machining ECM and electrical dis-charge EDM, presented a good solution, because an aircraft is currently assembled between 2 to 8 months, depending on the size and the commercial nature, that means, assembly of 300 to 1000 Aircrafts per year. While the global automotive sector manufactures 80 a million vehicles per year, according to 2018 records referred to in OICA International Organization of Motor Vehicle Manufacturers [6]. This is where new high-performance materials challenge manufacturing processes as "cutting, forming-stamping, bonding and machining" play an important role. Consequently, manufacturing engineers are challenged to find viable highly inno-vative solutions.

The simultaneous hybrid ED + EC technologies not exist yet commercially for industrial use, being in development the machining by electrical discharge assisted with simultaneous electrochemical pulses ED + PEC or named Pulsed ECDM, the first of thermal nature of plasma-ionic type and the second electro ionic of chemical nature; is possible will be offered in this decade by the original equipment manufacturing houses (OEMs), for industrial applications.

A recent study called "Special Machine Tool Market by Research" [7], reveals that approximately 78,000 units were sold of special machine tools for the manufacturing of the cell-laser type, plasma cutting machines, EDM cutting machines, cutting machines ECM, Water-Jet Cutting machines, and CMM (Coor-dinate Measuring Machines), for a world market size with revenues of 9.6 billion dollars. Of which 22% of these units are made up of laser/plasma cutting, 45% are from the EDM process, 6.5% from units sold in ECM, and just over 15% for "Water-Jet" cutting technologies, and 11.5% remaining in coordinate measurement machines (CMM). Laser/plasma cutting tools and EDM are the processes of greater demand; it is a market that has not yet reached maturity with a growth of 7.8% per annum CAGR (Compound Annual Growth Rate).

To improve the application of the cutting process by electro-discharge EDM, it has been proposed to assist it with PEC pulsed electrochemistry, this class of processes is known as hybrid ECDM (Electrochemical discharge machining) or ECSM (Electro-chemical spark machining) as reported [8]. There are two categories of hybrid machining processes (HMP) as shown in **Figure 1**. Relationships of binary and ternary hybrids are based on their physical nature to carry energy (mechanical, chemical, and thermal). The first category of HMP's is that all its constituent processes are directly involved in the removal of material. The second category of assisted HMP's is made up of processes in which the only one of the constituent processes directly remove the material, while the others are only assisting the removal process, changing the machining conditions in an appropriate direction, improving the machining conditions. Some processes such as plastic flow, mechanical abrasion, heating, melting, evaporation, dissolution, manage to change the physicochemical conditions of the material of the workpiece during a machining process [9].

The application characteristics of hybrid processes are considerably different from the corresponding characteristics of the "constituent" processes when these are applied separately. For example, it is established that the productivity of ECM electrochemical machining, when assisted with EDM electric discharge, is 10 to 50 times higher [10, 11].

## 2. Hybrid simultaneous ED/PEC drilling using double cathode

The pulsed electrochemical machining (PECM) on a simultaneous pulsed train of discharge plasma EDM, on the surface of a workpiece is named simultaneous ED + PEC drilling. Combination electro discharge and chemical machining in low-resistivity deionized water, has been investigated in the last decade to obtain a high material removal rate (MRR) and transfer energy to the workpiece [12].

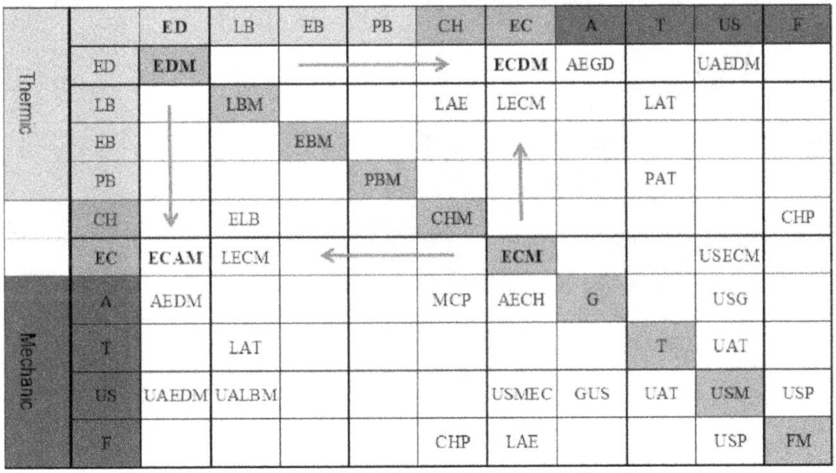

| | | ED | LB | EB | PB | CH | EC | A | T | US | F |
|---|---|---|---|---|---|---|---|---|---|---|---|
| Thermic | ED | EDM | | ⟶ | | ⟶ | ECDM | AEGD | | UAEDM | |
| | LB | | LBM | | | LAE | LECM | | LAT | | |
| | EB | | | EBM | | | ↑ | | | | |
| | PB | | | | PBM | | | | PAT | | |
| | CH | | ELB | | | CHM | | | | | CHP |
| | EC | ECAM | LECM | ⟵ | | | ECM | | | USECM | |
| Mechanic | A | AEDM | | | | MCP | AECH | G | | USG | |
| | T | | LAT | | | | | | T | UAT | |
| | US | UAEDM | UALBM | | | | USMEC | GUS | UAT | USM | USP |
| | F | | | | | CHP | LAE | | | USP | FM |

**Figure 1.**
*Advanced Hybrid Machining Processes (HMP) El -Hofy [1].*

## 2.1 Configuration double cathode for hybrid S-ED/PEC drilling

The configuration of a hybrid S-ED/PEC process in a semi-dielectric medium comes from a base EDM system, as a scheme is shown in **Figure 2**. The EDM equipment was complemented with two feed inputs: (i) dielectric deionized water (DW) and (ii) low resistivity deionized water (LR-DW), thus causing simultaneous EDM and ECM operating conditions in different regions of a Dual Cathode system. For EDM it is possible to use a graphite electrode in the form of an external head (first cathode), and for ECM an electrode composed of a set of 12 pins mounted on a bronze ring inside the graphite head, the external electrode presents an arrangement of channels that allows movement of the semi-electrical fluid and ionic transfer, these electrodes are arranged in such a way that they connect with an arrow that allows a rotation between 1200 to 1600 rpm, the head is electrically isolated and both samples have electrical continuity fed by a VDC source external pulsed.

In this configuration, a fluid can be fed in two ways. The EDM case feeds: a central inlet deionized water flow V1, through the system of cathodes arranged in a shape concentric with the head of the system. The head speed parameter Vz in the "z" axis constant at $0.5 \pm 0.05$ µm/s, up to a penetration height of 3 mm (H). On the other hand, stop the S-ED/PEC process, the equipment is configured using an ii) external input flow to the electrode with deionized water of low resistivity LR-DW and it is switched with the second flow V2 to the interior changing the deionized water DW by low resistivity water 0.5 MΩcm LR-DW. Sodium bromide salt (NaBr) in 1.23 ppm TDS was used to adjust the resistivity to 0.5 to $2.5 \pm 0.01$ MΩcm. NaBr has the ability to solvate ions and, therefore, show a constant electrical conductivity behavior for high temperatures reported by [13].

## 2.2 Theoretical model S-ED/PEC

The main characteristic of the proposed S-ED/ PEC (Simultaneous Electro Discharge/Pulsed Electrochemical), allows a significant increase in the efficiency of MRR, and a significant reduction in surface roughness, thus providing a better surface finish. It is well known that the EDM process contributes significantly to MRR, as it produces deep layer of heat affected zone (HAZ). While that the main contribution of ECM process as consequence of combining, is the removal HAZ layers allowed roughness less than 2.5 µm (Ra), as reported Levy GN and Maggi F (1990) [14].

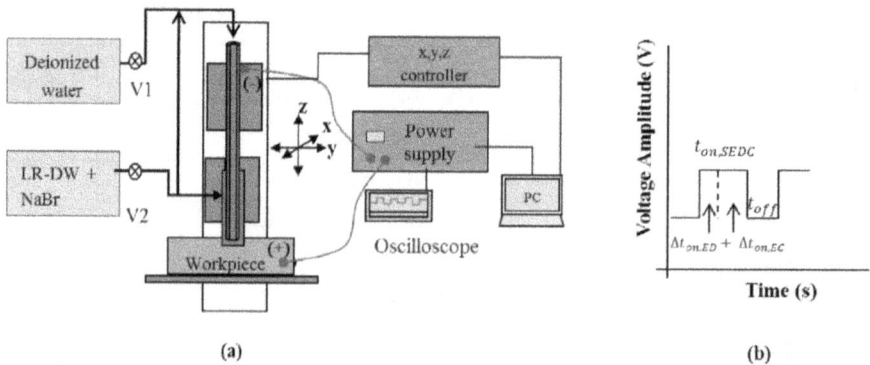

(a)  (b)

**Figure 2.**
*Experimental scheme of (a) EDM River 300 Cell - Instrumented, (b) ED / PEC open circuit voltage pulsed signal [15].*

They conducted a study on W-EDM for the machining of high-quality heat-treated alloy steels. They reported that the HAZ and the solidified layer reach 25 μm. Meanwhile, heat-affected zone with a white layer of approximately 10 μm with high hardness are reported [15]. On the other hand, the novel material removal process of high efficiency by blast erosion arc machining (BEAM), has an extremely higher material removal rate in relation to traditional EDM. However, the thickness of the HAZ caused by BEAM is close to 200 μm. Although it is known that the depth of the HAZ and the re-solidified layer is proportional to the amount of energy used.

Machining by S-ED /PEC, under dielectric conditions of low resistivity results in a phenomenon of physic-chemical activation, on the surface of the material that allows an exchange of advantages of the constituent that substantially improve the removal of material at high speeds with minimal thermal impact. Although, the contributions of the ED and PEC processes are not fully explained in the literature. In this research work, a mathematical model for ED/EC simultaneous drilling is proposed to determine the removal rate and the proportion of energy transferred to the workpiece under a new theoretical model, as well as to minimize the white layer effect to determine the contribution of each process in drilling holes in a High Strength Steels (HSS).

The combination of two phenomena, known as: (i) electro-thermal discharge and, (ii) electro-ionic dissolution, in simultaneous ED/PEC, increases the speed of material removal through chemical and physical activation of the metal surface, due to the exchange of advantages. A conceptual scheme of the removal mechanism for drilling by EDM and comparatively by S-ED/PEC is presented in **Figures 3** and **4**. The initial surface condition for $t_{i=0} = 0$, the volume elimination is VR = 0, as presented in **Figure 3(a)** for ED. After the first discharge condition $t_{i+1} = t_{on,ED}$ the volume $VR_1 = VR_s$ (elimination of volume by sparks), as indicated in **Figure 3(b)**. It is known that for every spark produced for EDM, this generate high roughness. In **Figures 3(c)** and **(d)**, the second discharge condition is $t_{i+2} = 2t_{on,ED}$, and the second volume is defined as: $VR_2 = \varphi VR_s$, where the fraction φ <1, $VR_2 < VR_1$ and the volume release is not equal to the first download.

On the other hand, the simultaneous ED / PEC drilling for the ED condition reveals that the surface initially for $t_{i=0} = 0$, and VR = 0, as shown in **Figure 4(a)**. After the first discharge for the first time stage $\Delta t_{on,ED}$, the volume removal is $VR_{s1} = VR_s$ as **Figure 4(b)**. In the second stage $t_{i+2} = \Delta t_{on,EC}$ there is a change in the elimination of EC by dissolution of the $Fe^{+2}$ ion, then the volume $VR_2 = VR_d$ as **Figure 4(c)** and **(d)**, with a lower surface roughness. Then, the value of the volume fraction is finally φ = 1, and therefore $VR_{s2} = VR_{s1}$, for each S-ED/PEC cycle. Following the EC condition, the surface shows better behavior due to the amount of material removed by the electric discharge process. This lower degree of roughness is produced by ion exchange during the electrochemical dissolution stage of the workpiece.

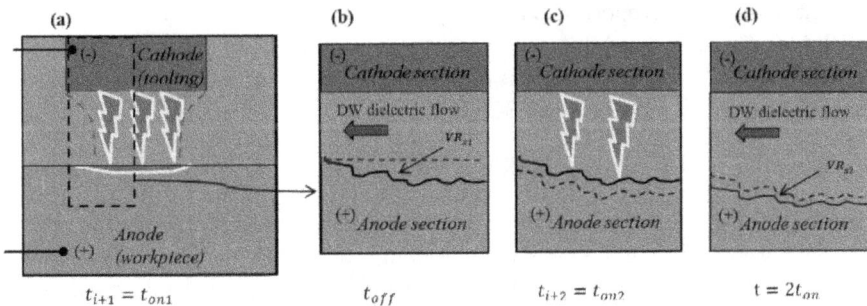

**Figure 3.**
*Diagram for EDM removal mechanism. (a) Initial discharge, (b) EDM first cycle $VR_{s1}$, (c) EDM second cycle and (d) Surface removal $VR_{s2} = VR_{s1}$ for φ <1 [16].*

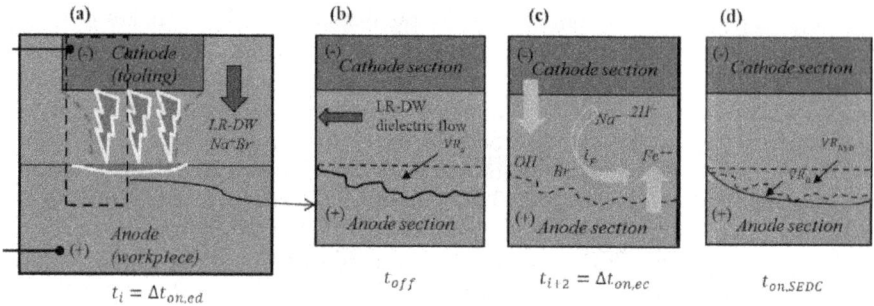

**Figure 4.**
S-ED/PEC drilling (a) initial discharge ED, (b) removal cycle $VR_s$ for ED, (c) volume $VR_d$ of the removed cycle for EC, and (d) S-ED/PEC total cycle volume $VR_{hyb} = VR_s + VR_d$ for $\varphi = 1$ [16].

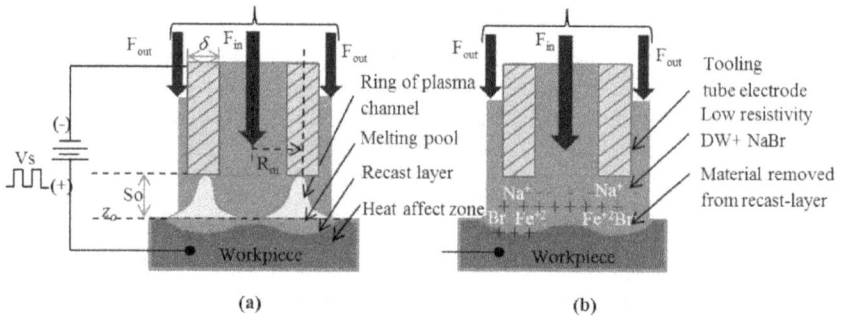

**Figure 5.**
Tool-workpiece scheme for S-ED/PEC drilling under two divided pulse conditions as (a) $\Delta t_{on1}$ for ED and (b) $\Delta t_{on2}$ for EC [16].

## 2.3 Mathematical model of simultaneous ED/PEC assistance

**Figure 5** shows the tool-workpiece scheme for the S-ED/PEC hybrid. There are irregular layers in the order of micro thicknesses located in the active region, due to the electric discharge of the EDM, as seen in **Figure 5(a)**. There are three main layers on the $z_o$ surface after discharge. These include (i) melt layer, (ii) the white layer is a remelting layer, and (iii) the heat affected zone.

**Figure 5(b)** shows the second mechanism of material removal, which involves the release of atomic layers, which is due to the electro-ionic dissolution process of the ECM. Therefore, the material removal rate per unit area $MRR^p$ $[mm^3 \cdot m^{-2} \cdot s^{-1}]$ per unit time $t$ [s] is proportional to the cycle energy function $E(t)$ $[J \cdot cycle^{-1}]$, which is required for machining. As a result, the machinability constant $K_m$ $[mm^3 \cdot m^{-2} \cdot J^{-1}]$ of this system is obtained the Eq. (1), as was written in the research theoretical model S-ED/PEC [16].

$$MRR^p * t = K_m * E(t) \tag{1}$$

Eq. (2) is used to calculate $MRR_{hyb}$ of low resistivity S-ED/PEC drilling in deionized water (LR-DW), where the pulsed duration takes two relevant conditions, namely (i) $\Delta t_{on,ED}$ for the ED condition and (ii) $\Delta t_{on,EC}$ for the EC condition. Therefore, $t_{on} = \Delta t_{on,ED} + \Delta t_{on,EC}$ is set by the simultaneous condition.

$$MRR_{hyb} t_c = \left( MRR_s^p \Delta t_{on,ED} + MRR_d^p \Delta t_{on,EC} \right) A_T \tag{2}$$

Substituting the respective expressions of Eq. (1), in Eq. (2), the resulting volume of material extracted during S-ED/PEC drilling could be calculated by means of Eq. (3), where $MRR_{hyb}$ is the velocity volume elimination $MRR_{hyb}$, and $K_s$, $K_d$ are the machinability constants of the ED and EC subsystems, respectively.

$$MRR_{hyb} = \left[ \left( \frac{1}{t_c} K_s * E_s(t) \right) + \left( \frac{1}{t_c} K_d * E_d(t) \right) \right] A_T \qquad (3)$$

For which, $A_T$ (mm²) is the cross-sectional area of the tubular electrode and is represented by Eq. (4).

$$A_T = 2\pi * R_m * \delta \qquad (4)$$

Considering $R_m$ (μm) as the average radius of the tool and $\delta$ (μm) is the thickness of the tool.

The amount of energy for each subsystem, $E_s(t)$ and $E_d(t)$, for a hybrid S-ED/PEC condition is represented by the terms of Eq. (3), which can be solved using an experimental pulsed electrical signal. It is known that the expression $E_s(t)$ for EDM can be defined by Eq. (5).

$$E_s = \int_0^{ton} I_e * V_s \, dt \qquad (5)$$

Where:

$$I_e = f_e * I_p \qquad (6)$$

While $E_d(t)$ for ECM it is possible to represent it by Eq. (7) [16].

$$E_d = \int_0^{ton} V_a * i_{o,Fe} * exp\left( \eta_{Fe} * \beta^{-1}_{a,Fe} \right) dt \qquad (7)$$

Where $V_a$ is the anodic voltage (volts) and $I_F$ is the Faraday current (A) that can be determined using TAFEL Eq. (8).

$$I_F = i_{o,Fe} * e^{\left( \propto \frac{zf}{RT} \right) * \eta_{Fe}} dt \qquad (8)$$

$i_{o,Fe}$ is the ion-exchange current (A) and $\eta_{Fe} = \left( \phi_{a,Fe} - \phi^0 \right)$ is the overpotential (volts) where $\phi_{a,Fe}$ is the anodic potential and $\phi^0$ is the equilibrium potential of the redox reaction with a value of −510 mV for Fe → Fe+2 + 2e.

Was considered the argument $\beta_{a,Fe} = \frac{RT}{\alpha z F}$ where $\beta_{a,Fe}$ (Volts⁻¹) represent the TAFEL anodic constant. The value of $\beta_{a,Fe}$ can be estimated assuming the following values. A symmetry coefficient of polarization $\alpha$ is equivalent to ½, ideal gas constant $R = 8.314$ JK⁻¹ mol⁻¹, Faraday's constant $F = 96487$ C mol⁻¹, the electron exchange number $z$ for iron case Fe → Fe+2 + 2e, $z = 2$, and $T$ is the interphase temperature assumed Temperature of room. Then, this can be estimated as $\beta^{-1} = 31.97 \, mV$ as was reported by Winston [17], for a cell S-ED/PEC.

## 2.4 Simultaneous ED/PEC electro-ionic model

It is possible to determine the fraction contributions of each process for the S-ED/PEC hybrid, where $\psi_s$ (9) is the contribution of the ED fraction and $\psi_d$ (10) is the contribution of the EC fraction. The hybrid process can be expressed as $\psi_s + \psi_d = 1$. Therefore:

$$\psi_s = \frac{VR_s}{VR_s + VR_d} = \frac{K_sE_s}{K_sE_s + K_dE_d} \tag{9}$$

The expression EC fraction can be written as:

$$\psi_d = \frac{K_dE_d}{K_sE_s + K_dE_d} \tag{10}$$

To determine the machinability constant of the simultaneous ED/PEC drilling, Eqs. (3) and (11) are used and was obtained $K_{hyb}$, which is defined as Eq. (12):

$$MRR^p * t_c = K_s * E_s + K_d * E_d = K_h * (E_s + E_d) \tag{11}$$

$$K_{hyb} = \frac{K_sE_s + K_dE_d}{E_s + E_d} \tag{12}$$

A simplification of Eq. (3) in terms of machinability constants for the hybrid $K_{hyb}$, the contribution fraction $\psi_s$ of the ED process in Eq. (9) and the final contribution fraction $\psi_d$ of the EC process in Eq. (10) results in Eq. (13), and it is solved by obtaining the experimental constants for the LR-DW condition of the simultaneous ED /PEC drilling. A resistivity range, in the order of 0.5 to 2.5 MΩcm is used to obtain the hybrid machinability constant $K_h$, where the pulsed activation time of each constituent ED and EC are fractions of the active time condition of the ED /PEC simultaneous process.

$$MRR_{hyb} = K_h * A_T \left[ \left[ \frac{1}{t_c} * \int_0^{ton,ED} (Ie * V )dt \right] + \left[ \frac{1}{t_c} \int_0^{ton,EC} \left[ V_a * i_o * exp\left( \eta_{Fe} * \beta^{-1}_{a,Fe} \right) \right] dt \right] \right] \tag{13}$$

A methodology was developed that allowed to reproduce the basic processes ECM, EDM and the hybrid ED/PEC considering the drilling process in the test materials, HSLA high-strength steel, in thicknesses of 9.5 mm, to compare the speed of material removal (MRR). Preparation of 32 samples of 9 x 12 x 35 mm of HSS-550 (HSLA). A state of the art search was carried out in the hybrid electro-ionic-based ECM and electroplasma-based EDM processes, which supports the knowledge base for the development of the simultaneous hybrid ECM + EDM model.

## 2.5 Experimental parameters for experimental drilling EDM, ECM and S-ED/PEC

To validate the proposed model with the observation, measurement and comparison of the electro-thermal/electro-ionic effect on the workpiece in High Strength Steel Low Alloy (HSLA), the microstructure at the cutting front was evaluated for each condition EDM, ECM, and compared against the simultaneous hybrid S-ED / PEC. The drilling parameters of each system were established according to the theoretical framework developed, for the different EDM, ECM and S-ED / PEC processes in different media as shown in **Table 1**. Similarly, the parameters for the ECM, EDM and S-ED/PEC processes were defined as shown in **Table 2**, which were used in the proposed experiment designs for each route according to the methodology. The parameters were determined based on the

| Machining process | Medium/electrode | Type | Magnitude | Voltage DC |
|---|---|---|---|---|
| ECM | Electrolyte | Acid Solution [Na2NO3 + H2SO] | ~10 Ω.cm | Continuous |
| EDM-OILD | OILD Dielectric | Oil | ~ 10 MΩ.cm | Pulsed |
| EDM-DW | DW Dielectric | [H2O] ~Deionizade | 2.5 MΩ.cm | Pulsed |
| S-ED/PEC | Semidielectric (LR-DW) | [H2O] + [NaBr] | 0.5 a 2.5 MΩ.cm | Pulsed |
| ECM | Electrode | Tube SS-304 | ⌀ 2.5 mm δ 250 μm | Pulsed |
| S-ED/PEC | Electrode | Tube Cu-Sn | ⌀ 500 a 1000 μm δ 100 a 150 μm | Pulsed |

**Table 1.**
*Specification for ECM, EDM and S-ED/PEC machining [16].*

| Parameters | Simbology | Values | Units | Condition |
|---|---|---|---|---|
| Voltage | VS1,VS3 | 12,30,45 | Volts | Variable |
| Gap Voltage | VG | 45,60 | Volts | Variable |
| Current | $I_c$ | 10, 15, 25 | Ampers | Variable |
| Pulsed time | $t_{on}$ | 12,20,28 | μS | Variable |
| Cycle Duty | DC | 0.3,0.5,0.7 | % | Variable |
| Frecuence | $\tau^{-1}$ | 25000 | Hz | Constante |
| Resolution | $V_z$ | 1,5 | μm | Constante |
| Cutting Deep | H | 3.0, 3.5 | mm | Constante |

**Table 2.**
*Parameters for ECM, EDM and S-ED/PEC machining [16].*

electrical range Vs, discharge start voltage from 30 to 45 Vdc, and electrochemical resistivity, 0.1 to 0.5 MΩ · cm, which allows a stable and reproducible ED + EC hybridization [18].

## 3. Discussion and results

The profile of the electrical signals for EDM and ED / PEC simultaneous drilling is presented in **Figure 6**. In particular, a typical electrical signal for the EDM process is shown in aqueous medium in deionized water at 2.5 MΩ · cm. **Figure 6(a)** shows the results of the energy transferred $E_s(t)$ during the ED cycle considering the area under the curve of the relationship Voltage Vs Current. **Figure 6(b)** reveals the results of the pulsed electrical signal, which was modified using a resistivity close to 0.5 MΩ · cm in deionized water adjusting the solution with NaBr ions. The result is a simultaneous ED/PEC behavior where the current wave was monitored with a decrease in the Faraday current. Where a division of the electrical signal is seen in two energy regions: in the first section of the current profile, up to where the current plateau ends, the energy magnitude $E_s(t)$ is obtained, which is determined by the process ED; while the current decays exponentially to a value close to the nominal current $I_o$, it is possible to obtain a second region of energy $E_d(t)$, which results from the EC process, for each pulse or duty cycle with a duration of 28 μs.

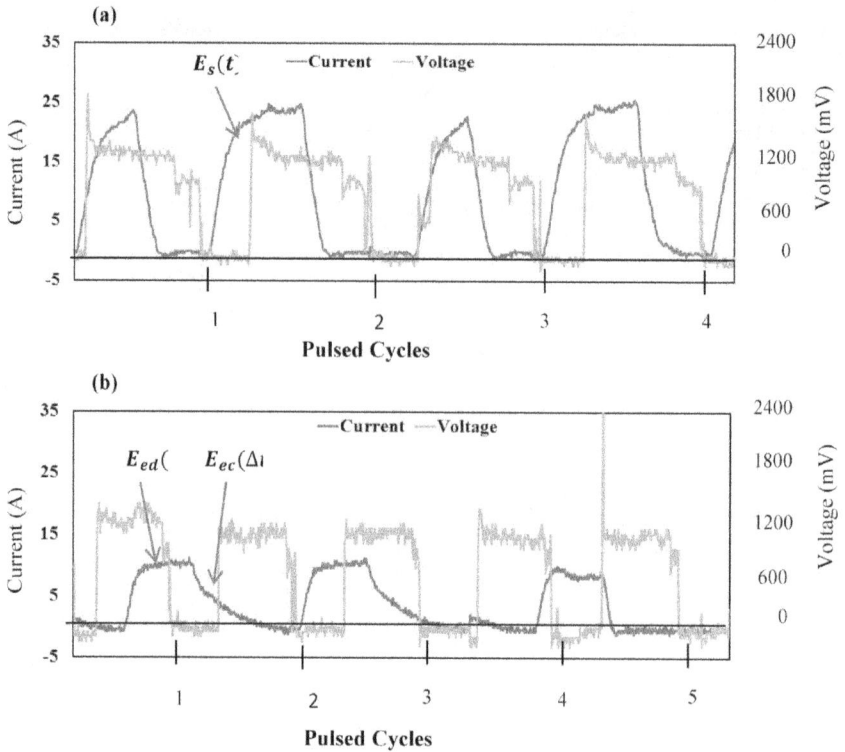

**Figure 6.**
*Showed electrical signal performance using a 25 kHz pulsed with DC 70% during (a) EDM and (b) hybrid ED/PEC [16].*

The values of energy transferred and machinability constant shown in **Table 3**, for the theoretical model $MRR_{hyb}$ represented by Eq. (13), are determined under two power conditions, 25A and 15A. Eqs. (5), (7) and (12) obtain the values of the transferred energy $E_s$, $E_d$ y $E_{hyb}$, while the machinability constants $K_s$, $K_d$, $K_h$ were obtained from Eqs. (3) and (12). Eqs. (9) and (10) estimate the energy fraction $\psi_s$ for ED and the energy fraction $\psi_d$ for EC, which contributes to S-ED/PEC machining.

The response of the theoretical model is shown in the MRR profiles for the ED/PEC and EDM processes. **Figure 7** shows the profiles of the increase in MRR for

| Magnitudes | High power (25 A/45 V) | Low power (15A/30 V) |
|---|---|---|
| $E_s\ [J \cdot cycle^{-1}]$ | 1125 | 450 |
| $E_d\ [J \cdot cycle^{-1}]$ | 45 | 27 |
| $E_{hyb}\ [J \cdot cycle^{-1}]$ | 1170 | 477 |
| $K_s\ [mm^3 \cdot mm^{-2} \cdot J^{-1}]$ | $4.342 \times 10^{-4}$ | $5.497 \times 10^{-4}$ |
| $K_d\ [mm^3 \cdot mm^{-2} \cdot J^{-1}]$ | $1.734 \times 10^{-5}$ | $2.180 \times 10^{-5}$ |
| $K_h\ [mm^3 \cdot mm^{-2} \cdot J^{-1}]$ | $5.070 \times 10^{-4}$ | $6.651 \times 10^{-4}$ |
| $\psi_s$ | 0.924 | 0.907 |
| $\psi_d$ | 0.075 | 0.092 |

**Table 3.**
*Transferred energy for S-ED/PEC machinability [16].*

EDM, in a response surface methodology (RSM) using Design-Expert® Software Version 10.0. As the pulse duration increases, the MRR increases as shown **Table 4**, a trend similar to that reported by Shabgard and Akhbari [19]. They report the effect of discharge current and pulse duration for EDM and ECDM, respectively. In this analysis, it was observed that when the current decays, it is necessary to increase the pulse duration, to maintain a constant MRR profile.

The resistivity analysis for MRR during ED /PEC is shown in **Figure 8**, with a LR-DW NaBr medium, in the range of 0.5 to 2.5 MΩm at 15A. **Figure 8(a)**, the response surface analysis shows the duration of the pulse versus the resistivity of the electrolyte. The MRR exhibits a slight increase in low resistivity near the 0.5 MΩ.cm condition at 12 μs. When the resistivity exceeds 1.5 MΩ.cm, the system works in the EDM condition, there is a transition point from the EC to ED condition. This is observed in **Figure 8(b)**, the response in the removal of MRR material as a function of the Voltage for different levels of resistivity of the medium, a low voltage region of the MRR profile is appreciated at 8 mm$^3$/min, where it reveals an inflection point, where the resistivity is greater than 2.0 MΩcm. This finding is consistent about the ED/EC transition that exists in the low resistivity medium for EC was reported by NGuyen et al. [20]. Under conditions similar to ECM where the system operates under low current and voltage conditions.

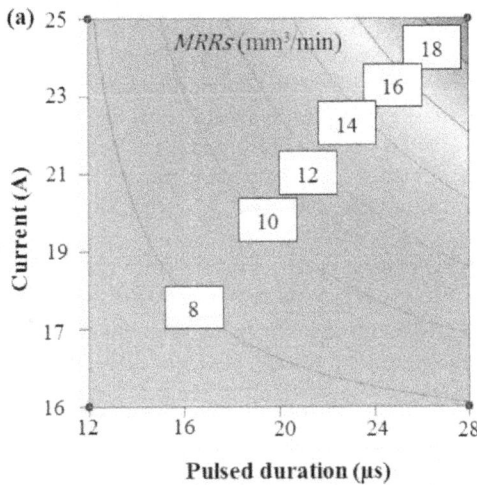

**Figure 7.**
*Response surface analysis for material removal rate MRR during EDM at 2.5 MΩcm/45 V [16].*

| Parameters | EDM | ECM | S-ED/PEC |
|---|---|---|---|
| MRRv (mm$^3$/min) | 20 a 22 | 2.8 a 3.4 | 23–28 |
| Over cutting (%) | <30 | 40 a 120 | <36 |
| White layer (μm) | 3.0 a 6.5 | 0 | 0.5 a1.5 |
| HAZ (μm) | 10 a 20 | 0 | <6.25 |
| Hardening HV | 320 a 380 | MB | <300 |
| Quality | Poor(microfails) | Excellent | Good |

**Table 4.**
*Comparison of EDM cutting processes ECM and S-ED/PEC for HSS Domex 550C [16].*

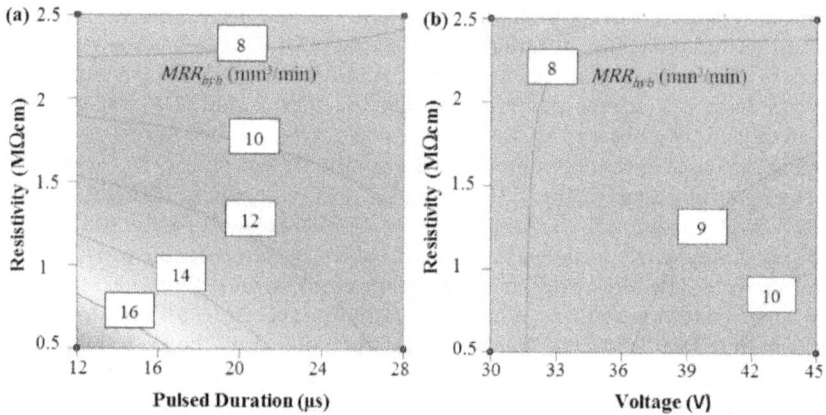

**Figure 8.**
*Response analysis of MRR_hyb for (a) ED/PEC at 15A, 45 V and (b) ED/PEC at 15A and pulse duration of 12 μs [16].*

**Figure 8(b)** presents the results of the simultaneous ED/PEC sensitivity analysis to estimate MRR applying the theoretical model using a NaBr LR-DW in a 0.5 MΩcm medium. The two regions are clearly separated in a current close to 19A and with a pulsatile duration of 20 μs at the pulsed voltage. The first region is defined as a low current of less than 20 μs and, since it is primarily electrochemical, it is in the region of the ECM. The second region is in the high current range, that is, above 20 μs, and because it is primarily an electrical discharge process, it is in the ED Region. Therefore, ED → ECD → PEC processes produce simultaneous ED/PEC, and when higher currents occur with low pulse durations, the passivation effect occurs in the ECM condition. In the same way, high pulse durations with lower currents result in a low MRR.

**Figure 9**, compares the experimental results with the theoretical model of $MRR_{hyb}$ for the EDM material removal rate and simultaneous ED/EC drilling. The $MRR_{hyb}$ is formed by two conditions, material removal of ED and EC simultaneously. When the pulse duration is increased, the effect in both cases is an increase in the rate of material removal, also determined that the effective sparks are determined by the duration of the pulsed signal and the analysis of the removal of

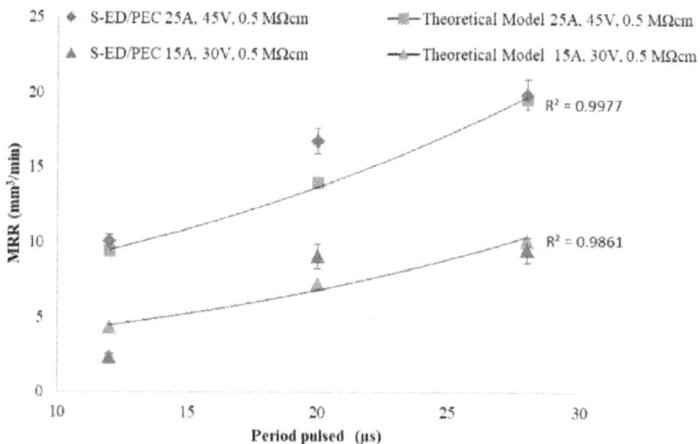

**Figure 9.**
*Comparison of the experimental MRR versus theoretical model for ED/PEC drilling in HSLA steel [16].*

material by discharge is significantly increased. However, the current has a greater effect on the MRR, because it generates an increase in the energy transferred to the workpiece, and associated with long periods of time produces a higher material removal rate and effect of surface roughness [21]. An MRR behavior at high current increases to 5 mm³/min by 10 μs. While the increase for a low discharge current is 2.5 mm³/min for 10 μs. This is consistent with the ECM process, which exhibits effective material removal rates in the range of 1.51 mm³/(A · min) to 2.13 mm³/(A · min) and that the speed increases as the density of current [22].

## 4. Conclusions

The analytical correlation of the mathematical model and the experimental results of the cutting face for simultaneous S-ED/PEC drilling using medium to low resistivity for drilling in HSLA material were established as follows.

1. A proposed parametric model for the simultaneous S-ED/PEC was developed against the experimental data to determine the MRR with an acceptable correlation close to 0.99, which was possible to obtain a machinability constant in the range of $5.07 \times 10^{-4}$ a $6.65 \times 10^{-4}$ (mm³ · mm$^{-2}$ · J$^{-1}$). The proportion of energy transferred that contributes to individual processes for simultaneous ED/PEC was estimated at $\psi_s = 0.9246$ for the ED fraction and $\psi_d = 0.0753$ for the EC fraction.

2. The effect of resistivity in the response sensitivity analysis, with respect to the material removal rate for simultaneous S-ED/PEC drilling, results in a change in direction of the MRR, specifically when the resistivity is less than 1.5 MΩ · cm. For the PECM contribution, at higher voltages with lower current, a slight increase in MRR occurs.

3. The effect of overheating in the EDM process increases the HAZ layer, so that it is six times greater than that obtained by the S-ED/PEC drilling of LR-DW. The results indicate that the contribution of PECM allowed the material removal mechanism to reduce the involvement in the microstructure through assisted dissolution.

## Acknowledgements

I want to give a huge thanks full to COMIMSA advanced manufacturing department and the researchers professors PhD Eduardo Hurtado, PhD Melvyn Alvarez and PhD Pedro Perez, all them my acknowledgments. Also, special mention to CONACYT Grant numbers 174568 – 2014.

## Abbreviations

| | |
|---|---|
| A | Symmetry factor |
| A | Abrasive |
| AECH | Abrasive Electro Chemical |
| AED | Abrasive Electrical Discharge |
| AEDM | Abrasive Electrical discharge machining |
| $\beta_{a,Fe}$ | TAFEL anodic constant $[mV^{-1}]$ |

| | |
|---|---|
| δ | Tooling thickness (μm) |
| $\eta_{Fe}$ | Anodic over potential of iron $[mV]$ |
| $\phi_{a,Fe}$ | Anodic potential of iron $[mV]$ |
| $\phi^0$ | Standard potential $[mV]$ |
| $i_{o,Fe}$ | Current of ion-exchange of iron $[A]$ |
| φ | Volume fraction by cycle |
| $\psi_s$ | Energy share factor ED |
| $\psi_d$ | Energy share factor EC |
| $\Delta t_{on,ED}$ | Active differential time for ED $[s]$ |
| $\Delta t_{on,EC}$ | Active differential time for EC $[s]$ |
| $A_T$ | Drilling area [mm2] |
| *CHM* | Chemical Machining |
| *CHP* | Chemical Pressure Jet |
| *DW* | Deionized water dielectric |
| *LR-DW* | Low-resistivity deionized water |
| *EBM* | Electro Beam Machining |
| *EC* | Electrochemical process |
| *ECM* | Electro chemical Machining |
| *ED* | Electro discharge process |
| *EDM* | Electro Discharge Machining |
| *ELB* | Electron Laser Beam |
| E(t) | Energy transferred $\left[ J \cdot cycle^{-1} \right]$ |
| $E_d$ | Energy transferred for EC $\left[ J \cdot cycle^{-1} \right]$ |
| $E_s$ | Energy transferred for ED $\left[ J \cdot cycle^{-1} \right]$ |
| $E_h$ | Energy transferred for SEDCM hybrid $\left[ J \cdot cycle^{-1} \right]$ |
| $f_e$ | Factor of equivalent current |
| F | Faraday constant $\left[ C \cdot mol^{-1} \right]$ |
| F | Fluid |
| $F_{in}$ | LR-DW/DW internal flow $\left[ mm^3 \cdot min^{-1} \right]$ |
| $F_{out}$ | LR-DW outside flow $\left[ mm^3 \cdot min^{-1} \right]$ |
| FM | Forces Magnetohydrodynamic |
| G | Grinding |
| *GUS* | Grinding Ultrasonic |
| $I_e$ | Current wave equivalent $[A]$ |
| $I_F$ | Faraday current $[A]$ |
| $I_p$ | Current of wave peak $[A]$ |
| $K_m$ | Machinability constant $\left[ mm^3 \cdot m^{-2} \cdot J^{-1} \right]$ |
| $K_d$ | Machinability constant for EC $\left[ mm^3 \cdot m^{-2} \cdot J^{-1} \right]$ |
| $K_s$ | Machinability constant for ED $\left[ mm^3 \cdot m^{-2} \cdot J^{-1} \right]$ |
| $K_h$ | Machinability constant SED/PEC $\left[ mm^3 \cdot m^{-2} \cdot J^{-1} \right]$ |
| LBM | Laser Beam Machining |
| MCP | Mechanical cutting plasm |
| $MRR^P$ | Planar material removal rate $\left[ mm^3 \cdot m^{-2} \cdot s^{-1} \right]$ |
| $MRR^P_d$ | Planar material removal rate EC $\left[ mm^3 \cdot m^{-2} \cdot s^{-1} \right]$ |
| $MRR^P_s$ | Planar removal rate for ED $\left[ mm^3 \cdot m^{-2} \cdot s^{-1} \right]$ |
| $MRR_s$ | Material removal rate for ED $\left[ mm^3 \cdot min^{-1} \right]$ |
| $MRR_{hyb}$ | Material removal rate for SEDCM $\left[ mm^3 \cdot min^{-1} \right]$ |
| *LAE* | Laser Abrasive Electrolyte |
| *LAT* | Laser Abrasive Turning |
| *LEM* | Laser Electrochemical Machining |

| PBM | Plasm Beam Machining |
| $R$ | Ideal gases constant $[J \cdot mol^{-1} \cdot K^{-1}]$ |
| RSM | Response Surface Methodology |
| $R_m$ | Average radius [μm] |
| S-ED/PEC | simultaneous ED/CM drilling |
| $S_o$ | Tooling-workpiece gap $[mm]$ |
| t | Time $[s]$ |
| $t_c$ | Cycle time $[s]$ |
| $t_i$ | Iterative time $[s]$ |
| $t_{on}$ | On-time $[s]$ |
| $t_{off}$ | Off-time $[s]$ |
| T | Temperature $[K]$ |
| T | Turning |
| Th | Thermal |
| UAEDM | Ultrasonic Abrasive Electrical Discharge Machining |
| UALBM | Ultrasonic Abrasive Laser Beam Machining |
| UAT | Ultrasonic-Abrasive Turning |
| USECM | USECM Ultrasonic Electrochemical Machining |
| USMEC | Ultrasonic Mechanical - Electrochemical |
| US | Ultrasonic |
| USG | Ultrasonic- Grinding |
| USP | Ultrasonic Pressure Jet |
| $V_a$ | Anodic voltage $[mV]$ |
| $V_s$ | Pulsed voltage of source $[V]$ |
| VR | Removal volume rate $[mm^3]$ |
| $VR_d$ | Removal volume rate for ED dissolution $[mm^3]$ |
| $VR_s$ | Removal volume rate for ED spark $[mm^3]$ |
| $z_0$ | Initial reference z-axis |
| $Z$ | Numbers of electrons exchange-redox |

## Author details

Jesus M. Orona-Hinojos[1,2]

1 Quimmco Innovation System, San Pedro Garza García, N.L., Mexico

2 Corporación Mexicana de Investigación en Materiales Quimmco Innovation System. San Pedro Garza García, N.L., México, USA

*Address all correspondence to: jesus.orona@quimmco.com

## IntechOpen

# References

[1] **El-Hofy H,** editorial McGraw-Hill. Advanced Machining Processes: Non Traditional and Hybrid Machining Processes. New York; 2005. p. 222-224.

[2] **Singh T and Dvivedi A:** A review on electrochemical discharge machining, process variants and their hybrid methods. International Journal of Machine Tools and Manufacture. 2016; 105:1-13.

[3] **Pawar P, Ballav R and Kumar:** A Revolutionary Developments in ECDM Process. An Overview. 4th International Conference on Materials Processing and Characterization Proceedings. 2015;2: 3188-3195.

[4] **Zhao W [et al.]** A novel high efficiency electrical erosion process: Blasting erosion arc machining. Blasting erosion arc machining: Congress Seventeenth International Symposium on Electro machining. Shanghai China. 2013;621-625.

[5] **David R. TaeuschClinton J. Wohlmuth:** Laser/EDM drilling manufacturing cell. EP1988 / 0299143 A1.

[6] **OICA** International Organization of Motor Vehicle Manufacturers [Internet]. 2018. Available from: https://www.oica.net/category/production-statistics/2018-statistics/

[7] **Frost- Sullivan,** editor. Special Machine Tool Market by Research: Global Machine Tools and Cutting Tools Market [Internet]. 2013. Available from: https://store.frost.com/global-machine-tools-and-cutting-tools-market.html#section1.

[8] **Bhondwe KL, Yadava V and Kathiresan G:** Finite element prediction of material removal rate due to electro-chemical spark machining . Int. Journal of Machine Tools & Manufacture. 2006; 46:1699-1706.

[9] **Kozak J, Gulbinowicz D and Gulbinowicz Z:** The mathematical modeling and computer simulation of pulse electrochemical micromachining. Simulation of pulse electrochemical micromachining. AIP Conference Proceedings. 2009;11:174-181.

[10] **Kozak J, Rajurkar K P and Makkar Y:** Study of Pulse Electrochemical Micromachining. Journal of Manufacturing Processes. 2004; 6:1-11.

[11] **Rajurkar KP, Sundaram MM and Malshe AP:** Review of Electrochemical and Electrodischarge Machining. The Seventeenth CIRP Conference on Electro Physical and Chemical Machining : (ISEM) Procedia CIRP. 201; 6:13-26.

[12] **Jesus M Orona-Hinojos:** Electrothermal modeling of the hybrid EDM+PECM machining process to determine the efficiency and rate removal material in HSLA steel [thesis]. Coahuila México: COMIMSA. 2018.

[13] **Bhargav P B [et al.]** Structural and electrical properties of pure and NaBr doped poly (vinyl alcohol) (PVA) polymer electrolyte films for solid state battery applications. Journal Ionics. 2007; 13: 441–446.

[14] **Levy GN and Maggi F** WED Machinability Comparison of Different Steel Grades. Annals of CIRP. 1990; 39: 183-186.

[15] **Xiao-lei Xu, Zhi-wei Yu and Yu-zhou Gao:** Micro-cracks on electro-discharge machined surface and the fatigue failure of a diesel engine injector. Engineering Failure Analysis. 2013;32: 124–133.

[16] **Jesus Orona-Hinojos JM, Álvarez-Vera M, Hurtado Delgado E, Macias Avila E,Granda Gutiérrez E, Pérez**

**Villanueva P.** Modeling Simultaneous ED/PEC In Low-Resistivity Deionized Water (LR-Dw) For Machining Hsla Steel Using High-Frequency Pulsed Source [Conference]: MemoriaElectro2017. Chihuahua, México : ISSN 1405-2172, 2017; 39. p. 120-128.

[17] **R.R. Winston**, Uhlig's Corrosion Handbook, 2nd Ed, Electrochemical Society Series. New York; 2000.

[18] **NGuyen MD, Rahman M and Wong YS:** Simultaneous micro-EDM and micro-ECM in low-resistivity deionized water. International Journal of Machine Tools and Manufacture. 2012; 54:55-65.

[19] **Shabgard S and Akhbari M:** An inverse heat conduction method to determinate the energy transferred to the work piece in EDM proces. Int Journal Adv Manuf Technology. 2016; 83:1037–1045.

[20] **NGuyen MD, Rahman M and Wong YS:** Transitions of micro-EDM/ SEDCM/ micro-ECM milling in low-resistivity deionized water. International Journal of Machine Tools and Manufacture. 2013; 69: 48-56.

[21] **Cheng-Kuang Y Chih-Ping C, Chao-Chuang M, Cheng WA , Jung-Chou H, Biing-Hwa Y:** Effect of surface roughness of tool electrode materials in ECDM performance. International Journal of Machine Tools and Manufacture. 2010;50:1088–1096.

[22] **Klocke F Zeis M, Klink A, Veselovac D:** Experimental research on the electrochemical machining of modern titanium- and nickel-based alloys for aero engine components. The Seventeenth CIRP Conference on Electro Physical and Chemical Machining (ISEM). 201;1: 368-372.

Chapter 4

# Technological Development of CNC Machine Tool for Machining Soft Materials

*Laura Tobon Ospina, Carlos Alberto Vergara Crismatt,*
*Juan David Arismendy Pulgarin, Sebastian Correa Zapata,*
*Yomin Estiven Jaramillo Munera, Jhon Edison Goez Mora,*
*Juan Camilo Londoño Lopera, John Sneyder Tamayo Zapata,*
*Carlos Eduardo Palacio Laverde and Edgar Mario Rico Mesa*

## Abstract

In recent years, the process carried out in the GACIPE research group is related to the development of the base technology of the manufacturing and metalworking industry. The machine tools that are vital for the consolidation and competitiveness of the industry in any country has been approached through two approaches: The design and construction of the new machine. In this aspect, the modeling of the structure and the displacements' parameterization allows defining the precision of the movements and the rational use of energy. The adaptation and repowering of a used machine. In this approach, the recovery and technological updating proposed to recover its performance, becoming an excellent alternative to improving and perfecting the production of a company. In both cases, the CNC milling machine tools are controlled by free software. The application proposed is mechanized in soft materials.

**Keywords:** CNC, modeling, repowering, free software, competitiveness

## 1. Introduction

Modern industry demands more precise and complex products that meet high-quality standards and protocols, using high-performance technological tools such as CNC machine tools. That according to Oxford Economics, the principal manufacturers and consumers worldwide are China, United States The United States, the European Economic Community, India, Brazil, and Japan. These countries have greater competitiveness in world industry [1]. India last decade has had accelerated growth in the consumption and the production of machine tools above 8 percent per year. That is a consequence of the incentives and policy change of the Indian government focused on the constitution of a park of machine tool development and the creation of a center of excellence in machine tools and production technologies [2]. In Germany, the area of machines and equipment is the second most important industrial sector. The country has become a technological engine that distinguishes

IntechOpen

itself as a high-tech nation worldwide [3]. The United States has distinguished itself as one of the leading exporters of machinery worldwide, being the principal supplier to China in recent years, demonstrating the big industrial capacity to develop machine tools. That equipment that par excellence is the basis for catapult the industrial production of a country [4]. In Latin American countries, MIPYMEs prioritize automation and technological development of machine tools, specifically in Colombia, MIPYMEs represent 94 percent of the industry, where most of the manufacturing processes of MIPYMEs in the metalworking sector are carried out in conventional machine tools [5].

## 2. CNC machine tools prototype

This section presents the design and development of a prototype of a CNC Milling Machine Tool in its mechanical, electrical, and electronic components and graphical interface.

### 2.1 Works modeling in CNC machine tools

The different publications scanned in the last ten years to analyze and study structural design in machine tools and cutting techniques. In [6] theoretical 3D model is presented that determines the instantaneous machining stresses generated by the cutting action, presenting the stress distributions in the body of the CNC machine. In [7] proposes a model about the instantaneous cutting forces in 5-axis machining systems per milling machine and identifies the machine's operating parameters to minimize machining error. In [8] explains the model to machine a prismatic workpiece whose objective is to evaluate each component of the machine's power absorption and its relationship with the cutting parameters. In [9] requires a methodology for optimizing the energy consumption to control the machine in the machining of the part is presented. That model is based on historical operating data. In [10] exposes method of defining parameters for NC and machining systems to allow energy efficiency with the least possible error in constructing the resulting part. In [11] proposes a mathematical model based on the cutter's numerical discretization, including the advance and inclination angles. This model is applied in certain typical conditions of a 5-axis machine. In [12] shows a method of cutting stability, the solution is proposed for milling type in multi-axis machines using the general projective geometry technique representing the cutting tool by an envelope of point clouds. In [13] presents a three-dimensional machining model that validates the conventional machining method taking into account the type of cut (uppercut milling, descending cut milling, milling style). Also, The machine used the thermal camera to observe the effective cut depth, shear force, and roughness. In [14] proposes a new cutting operation on a lathe called oriented cutting that consists of a cutter with a straight edge without inclination but oriented with an angle different from 90 degrees; tests show the influence of the edge of the tool on the forces of cut.

### 2.2 Structure of CNC prototype

In the development of the project, the modeling and mechanical design of the structure of the machine (**Figure 1**) is fundamental, taking into account the accessories of the machine such as servo motors, servo drives, sensors, guides, and sources that allow the movement of its axes speed and precision.

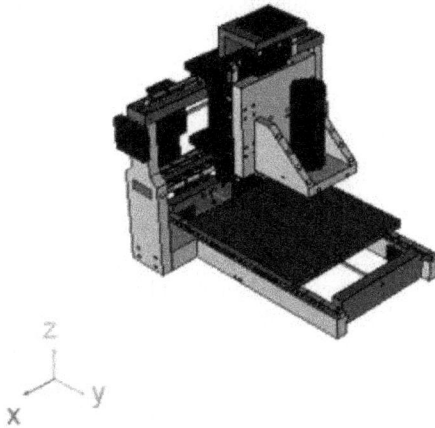

**Figure 1.**
*Isometric CNC milling machine.*

The structure has a frame, cross slide, table, bridge that supports articulation of the machine's three axes depending on to develop the pieces' machining. The table has a grooved surface on which the part to be shaped is held. The table is also supported by two carriages that allow the table's horizontal longitudinal movement on the transverse carriage (x-axis). The bridge is a cantilevered piece in the frame; there are also hardened and ground guides for vertical movement (y-axis). Some lunettes were supports of the horizontal movement axis (z-axis).

## 2.3 Development of the structure of the CNC machine

The machine was developed in a machining center from the model, taking into account the criteria and conditions obtained in the design and simulation of the structure to have robustness and precision in movement, in **Figure 2** you can see part of the real CNC milling machine armed consisting of the bridge, the cross carriage, the motor, and the guides. As shown in **Figure 2**, the machine's structure is configured as a vertical milling machine in which the table has movements in the x and y axes, and the motor with the spindle moves in the z-axis.

**Figure 2.**
*Real CNC machine.*

The machine structure was designed by modules and its construction in SAE 1020 plain steel material whose thickness is not less than 2.54 centimeter and with a manufacturing error of 10 micrometers.

## 2.4 Parameterization of movement of the CNC machine

The procedure for machining in CNC machine tools for milling machines with a fixed height table, horizontal or vertical spindle (**Figure 3**) has been sought, and the CNC machine is parameterized according to [15–17] with the following guide:

Verification of the straightness of the vertical and displacement of the spindle head. Tests are performed on a 155 mm stroke.

Measurements are made in the area tour, yielding the following results in **Table 1**.

The results were showed a variation lower than the admissible one (0.025 mm). This was measured in a length of 155 millimeters. A test carried out to verify the flatness of the table surface (**Figure 4**), in a 28x30 mm extension measurements, are made in the area tour, yielding the following results in **Table 1**. The practice was yielded the following results in different points of the said area, according to **Table 2**.

**Figure 3.**
*Displacement straightness check.*

| Displacement mm | Displacement error mm | Permissible error mm |
|---|---|---|
| 10 | 0.005 | 0.025 |
| 75.25 | 0.01 | 0.025 |
| 155 | 0.015 | 0.025 |

**Table 1.**
*Spindle vertical travel.*

**Figure 4.**
*Verification of the surface flatness of the table.*

| Coordinates mm | Real error mm | Permissible error mm |
|---|---|---|
| 0,0 | 0 | 0.05 |
| 0,150 | 0.03 | 0.05 |
| 0,280 | 0.025 | 0.05 |

**Table 2.**
*Flatness.*

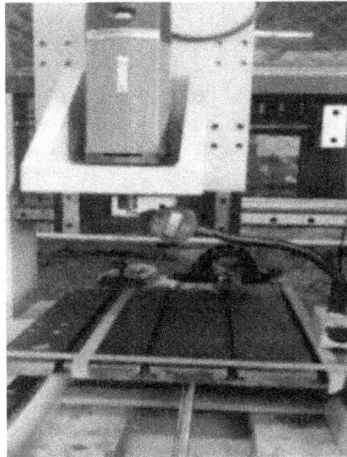

**Figure 5.**
*Checking the radial rotation of the inner core.*

The data was obtained in the test does not exceed the maximum allowable difference of 0.05 mm. The measurement of the radial jump of rotation of the inner core at the cone's exit is made (see **Figure 5**).

The test was resulted in the following measurements according to **Table 3**.

The data was obtained in **Table 3** indicate that it does not exceed the admissible difference of 0.02 mm during normal operation of the spindle.

| Proof | Measurement mm | Minimum error mm |
|-------|----------------|------------------|
| 1 | 0.02 | 0.02 |
| 2 | 0.01 | 0.02 |
| 3 | 0.02 | 0.02 |

**Table 3.**
*Measurement of radial jump in the spindle.*

## 3. CNC machine tools repowering

This section has presented the adaptation, reconditioning, and tuning of a used CNC machine tool.

### 3.1 Fine tunning CNC machine repowering

The experience of building a prototype of a CNC milling machine has been allowed to consider that it can apply the appropriate knowledge in used CNC machines that currently inoperative. Therefore, among the many technologically outdated CNC machines that SENA has, a CNC milling machine has been chosen EMCO brand didactic, used at the time to carry out academic activities with the institution's students. The conditions and characteristics of the machine are presented below in **Table 4**.

For the CNC machine to repower selected, its mechanical component (structure and mechanisms) must be functional and, as far as possible, close to its factory conditions, this allowed choosing the machine that is observed in **Figure 6**, The hardware of which was in good condition. However, some mechanical components were missing, and the electronic, electrical, and software module obsolescence.

The machine of **Figure 6** has been thoroughly inspected to define the roadmap for the recovery of the machine; therefore, the problems detected are presented below:

- The transmission system of movement towards each of the axes is carried out using a toothed belt and two pinions with a mechanical ratio of 2: 1. So, the torque is incremented but the speed of displacement is reduced.

- Elements such as gears, screws, supports, bearings, motors, guides, and bases were presented a considerable level of oxidation, preventing adequate movement and increasing the error tolerance in the axes' displacement by approximately 10 percent.

- During previous interventions, the ball screw of the "Y" axis and the movement transfer pinion of this same axis were not reintegrated into the machine, leaving it incomplete.

To mitigate mechanical difficulties, corrective maintenance related to the design, manufacture, and implementation of the missing parts is carried out;

| workspace in X, Y, Z [mm] | Spindle power at 2000 rpm [w] | Ball screw pitch [mm] | Accuracy of each axis [mm] |
|---------------------------|-------------------------------|-----------------------|----------------------------|
| 200,100,200 | 440 | 1.5 | 0.01 |

**Table 4.**
*Features repowered CNC machine.*

44

**Figure 6.**
*CNC machine repowering.*

subsequently, preventive maintenance is carried out to remove the oxide present and lubricate all the movement mechanisms; Once these corrections have been made, a mechanically adequate CNC milling machine is available to intervene in its entire electrical and electronic operating system.

## 3.2 Design and implementation of the electrical and electronic system of the repowered machine

Due to the service time of the milling machine, there are drawbacks such as shorts in the power circuit of the motors, failures in the control panel membranes, voltage drops in the power source due to problems in the voltage regulators; on the other hand, the software present in the machine is not compatible with the commercial applications that are currently used for this type of system, the main difficulty regarding compatibility is that the data access to the system is done using an obsolete technology (floppy disk 3/2), so the control card was designed and replace, the drivers and the user interface to recover their initial functions and even improve their operability.

A control and power board is implemented, see **Figure 7**, responsible for linking and optocoupler the drivers that control the motors of each of the axes with the instructions sent from the CNC software in the computer and sensors of end of career of the machine, stop of emergency and other elements that make up the system.

The machine's electronic system was made compatible with the LinuxCNC software; therefore, an optocoupler control interface (**Figure 7A**) has been designed and implemented connected to the parallel port of the pc. The signals obtained from the parallel port are input signals from sensors and output signals that enable actuating actuators. In the output signals, the power system of **Figure 7B** is used to control the actuators as servo motors and step motors responsible for moving the axes that position the milling cutter for machining development.

## 3.3 Parametrization of the repowered machine

For the execution of a machining production process, the CNC machine must meet a series of requirements related to the straightness in the movement of the axes, the flatness of the work table, and the parallelism concerning the cutting tool;

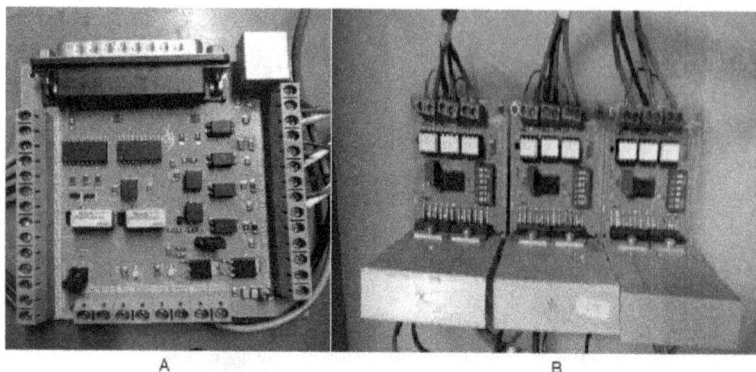

**Figure 7.**
*Electronic system (A) control target (B) power target.*

these parameters are necessary to bring the machine to work under the standards of machining parts and are verified by the following procedure:

To corroborate the straightness of the z-axis displacement, a dial gauge is positioned and measured concerning a square that is firstly parallel to the X-axis and later parallel to the Y-axis to verify that the displacement is strictly vertical and comply with the allowable error, which is 0.025 mm in a maximum length of 300 mm offset.

The table surface's flatness is measured with a level of precision where the tolerance cannot exceed 0.02 mm for every 300 mm of length between measurements.

The proceeding must guarantee the parallelism of the surface of the work table concerning the cutting tool, so the dial gauge is placed at its height (see **Figure 8**) and measurements are made along the length and width of the surface; the error cannot be greater than 0.025 mm for every 300 mm between measurements.

This procedure is applied to the machine and the results observed in **Table 2** are obtained, where the data are within the admissible limits to carry out a machining process.

### 3.4 Software selection and coupling with the CNC machine repowering

In the first instance, tests are carried out with the Mach3 software to verify the compatibility of the control card and the drivers with commercial software for the control of a CNC; the system presents good performance; However, the Mach3 is a demo version with limitations, for this reason, a search is carried out, finding some licensed software such as MasterCAM, SolidCAM, CAMWorks, among others, which were discarded because the license is very expensive, so, the best option has

**Figure 8.**
*Flatness teste.*

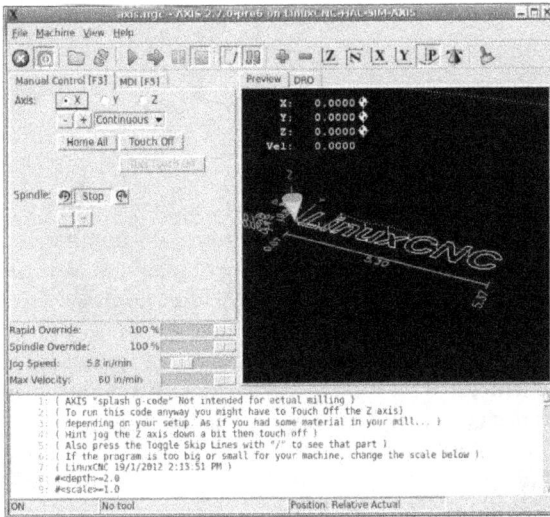

**Figure 9.**
*LinuxCNC interface.*

focused the search on open architecture software (Open Architecture CNC), finding that they handle great scalability, are freely distributed and present excellent compatibility with other software related to the manufacturing process and with the different CNC systems.

An application based on the Linux Ubuntu V10.04 operating system called LinuxCNC is selected, see **Figure 9**. The LinuxCNC has an architecture for the numerical control of CNC machines based on the RT-Linux kernel for the execution of instructions in real-time, with a capacity to control up to nine axes; Some works carried out with LinuxCNC demonstrate the capabilities and scope of the software such as those observed in [18–21], another application is observed in a robotic arm for to perform surgical procedures in [22], also shows precision and reliability in CNC systems as in [23].

## 4. Application used in the two machine tools (prototype and repowered)

To determine the response of machine tools, different types of parts have been machining. The tests have allowed knowing the real behavior of the CNC machines.

### 4.1 Machining tests to adjust the operations of CNC machines

During the machine set-up process, the relevant materials to be machined have been defined, taking into account the drive element's capacity (spindle) on both CNC machines. Therefore, although the machines' infrastructure allows machining in hard steels, there is a limitation in the spindle with steels. Thus, the tests will be carried out on 3000 series aluminum. This material has manganese as its main alloying material, which allows for good machinability.

To determine the optimal operating parameters; The speeds of movement of the axes and the revolutions per minute of the spindle are calculated using the following equations:

$$n = 1000(v)/d\pi \qquad (1)$$

$$f_r = f_z nZ \qquad (2)$$

$$T_m = (L + 2A)/f_r \qquad (3)$$

$$A = d/2 \qquad (4)$$

Where n = spindle speed in rev/min, v = cutting speed in m / min, d = diameter of the cutting tool in mm, $f_r$ = feed rate in mm/min, Z = number of cutting edges of the tool, $f_z$ = feed per edge in mm, $T_m$ = milling time in min, A = approach distance to hook the cutter to the material in mm, L = Milling length in mm. Note that Eqs. (2), (3) and (4) are only valid for milling operations.

The purpose of both the prototype machine and the repowered machine is to develop parts and structures for robotics applications through face milling. Initially, a piece numeric is designed simple, using a 4 mm flat milling cutter for roughing and a two-edged round milling cutter for finishing, both made of carbon steel, in this development of low complexity parts. The real piece is obtained with the design dimension proposed in the CAD (see **Figure 10**).

**Figure 10** shows the machining process of a numerical figure at various times from the beginning of **Figure 10A**, during **Figure 10B** and C, and the end of **Figure 10D**. This machining test was applied with the following machining operation parameters (see **Table 6**).

This information is obtained from several tests developed with the same design seeking to optimize the variables used in the LINUXCNC to achieve the materialization of the CAD (see **Figure 11**).

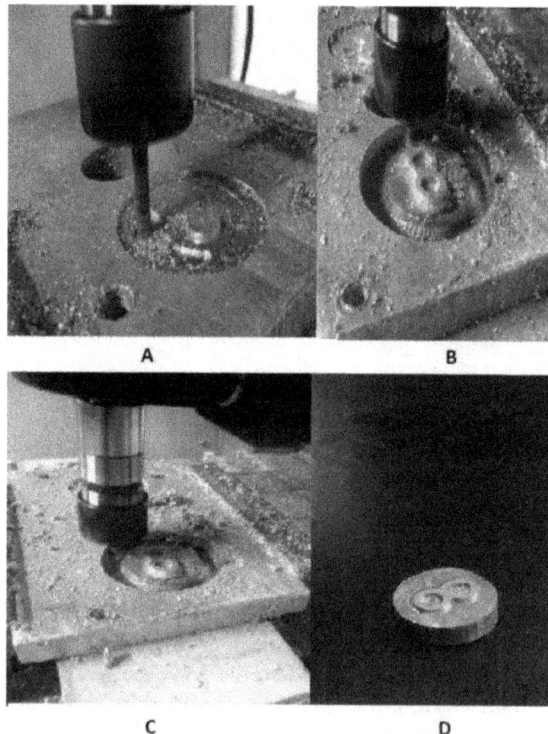

**Figure 10.**
*Alphanumeric part machining process (A) initial machined part. (B) Intermediate part machining (C) intermediate machining of the part figure (C) intermediate machining of the part contour (D) final machining of the part.*

| Description | Distance between measures [mm] | | | |
|---|---|---|---|---|
| | Ref | 100 | 200 | 300 |
| Z axis straightness with respect to x | 0 | 0.004 | 0.012 | 0.019 |
| Z axis straightness with respect to y | 0 | 0.002 | 0.004 | 0.007 |
| X-axis clamping table flatness | 0 | 0.006 | 0.011 | 0.015 |
| Y-axis clamping table flatness | 0 | 0.002 | 0.01 | 0.018 |
| Parallelism with respect to the x axis tool | 0 | 0.003 | 0.007 | 0.01 |
| Parallelism with respect to the y axis tool | 0 | 0.004 | 0.007 | 0.009 |

**Table 5.**
*Test scores.*

| Description | Process characteristics | | | |
|---|---|---|---|---|
| | n [rev/min] | $f_r$[mm/min] | $T_m$[min] | v [m/min] |
| Roughing of material | 6366 | 63.66 | 3.01 | 80 |
| Surface Finishing | 7957 | 15.9 | 31.44 | 100 |

**Table 6.**
*Fine tunning.*

In **Figure 11A**, shows the configuration of the axes. Parameters are defined as steps per revolution of the machine, the relationship between the motor shaft and the ball screw, the screw pitch, the speed and acceleration of the axis calculated based on the basic configuration information (**Figure 11B**). The axis length parameters and selection of the physical and virtual location of the home are configured. This menu is verified the entered configuration is correct by a test. In the basic information window of the machine, enter the name of the machine, the number of

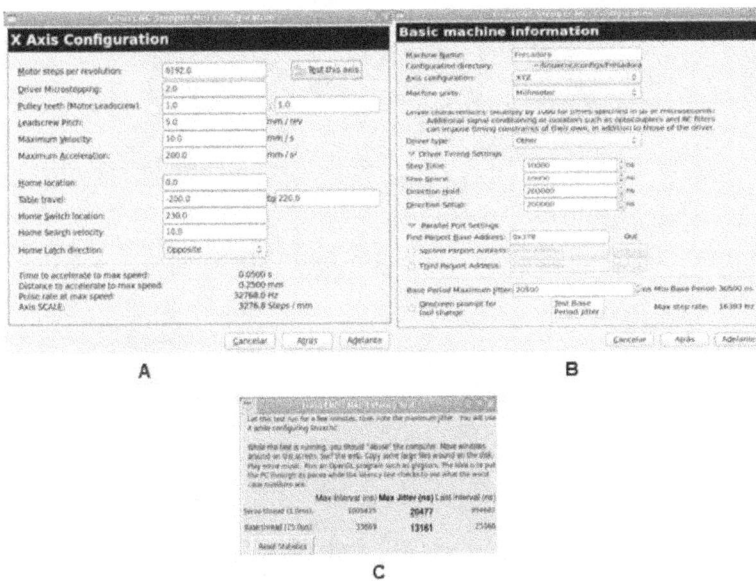

**Figure 11.**
*LinuxCNC parameters (A) axes configuration. (B) Machine information. (C) Latency test.*

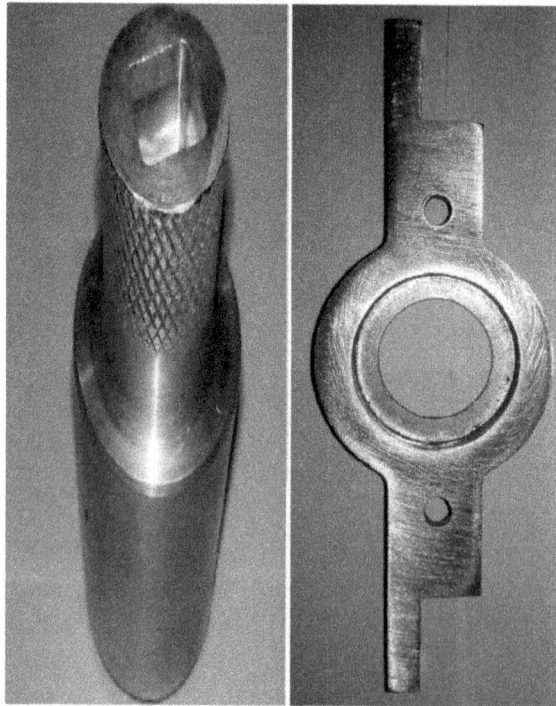

A                    B

C

**Figure 12.**
*Machined parts (A) motor shaft. (B) Shaft bracket. (C) Platform.*

| Operation | Piece | | |
|---|---|---|---|
| | Figure 12A | Figure 12B | Figure 12C |
| Planned | no | yes | yes |
| Squared | yes | yes | no |
| Contoured | no | no | yes |
| Boxed | yes | yes | no |

**Table 7.**
*Operations used in the milling of robotic parts.*

axes, and the units with which you want to work; To select the type of driver, the software provides some brands of manufacturers that have preset values; however, Linuxcnc allows the selection of a general type of driver and recommends standard values in which most of the drivers work properly. To find the system's maximum response time, the latency test is used (**Figure 11C**), where a test is verified the response capacity of the PC intended to operate the machine.

Subsequently, structural parts of a mobile robot were made for a real application in the health sector. The proposed parts have a more complex and larger structure; Therefore, 6 mm flat mills are used in cutting, and four-edged round milling cutters are used in the finishing. The machined parts fulfill the function of making traction and generating stability in the robot. According to what was projected in the CAD design, the real components are assembled in the robot structure without any difficulty and fully comply with the robotic system's function (see **Figure 12**).

The pieces of **Figure 12** are produced using machining parameters of **Table 6**, and **Table 7** shows some operations of machined pieces.

**Table 7** is about machining pieces of **Figure 12**, incorporated in the programming for prototype and re-powered CNC milling machine. In the milling of a piece is normal to find many types of machining operations. The piece more complex, according to **Table 7**, is the piece describes in **Figure 12B**. So the piece of **Figure 12B** applied various operations such as Planned, Squared, and Boxed.

## 5. Conclusions

In the development of the modeling of a CNC machine proposed in the works found [6, 8, 11] has been detected for the most part that energy optimization has started from a robust and functional structure as a fundamental premise. This aspect has been taken into account in this project in the design criteria used in the configuration of the structure of the CNC machine type milling machine.

In the construction and commissioning of the CNC machine have been considered in some works [7, 9, 10], the parameterization and recognition of the functionality as a fundamental development, so, three aspects have been put into practice in the present project: the straightness of the displacement of the coordinate axes, the flatness of the table surface and the radial jump of rotation of the inner core at the exit of the spindle. In the results of prototype and repowering was observed (see **Tables 1–3**) a good performance.

According to **Table 5**, the repowered CNC machine tool has been brought to an operating point similar to the factory's behavior. This allows the technological development to be exposed as a solution proposal with viable free software to update the CNC machine tools. So, the mechanisms and mechanical functionality of the machine are in good condition, and its recovery is viable.

The mechanization tests for both machine tools present the millimeter construction of robotics part in aluminum. The precision of prototype and repowered machines is achieved by determining the values of the machining parameters. These data are presented in **Table 6**. The pieces are possible to mechanize with a maximum area of 25 x 25 centimeters. The operations used in the machines'programming are Presented in **Table 7**, furthermore, the description is dimensioned the degree of complexity of each of the parts' machining.

## Author details

Laura Tobon Ospina*, Carlos Alberto Vergara Crismatt,
Juan David Arismendy Pulgarin, Sebastian Correa Zapata,
Yomin Estiven Jaramillo Munera, Jhon Edison Goez Mora,
Juan Camilo Londoño Lopera, John Sneyder Tamayo Zapata,
Carlos Eduardo Palacio Laverde and Edgar Mario Rico Mesa
SENA, Colombia

*Address all correspondence to: ltobon@sena.edu.co

IntechOpen

# References

[1] Oxford Economic, Global Machine Tool Outlook, Oxford University, pp 1–125, 2016.

[2] Radhika T R and Shripathi Kalluraya, Production trends in Indian machine tool industry, Int. Journal of Management and Development Studies vol 6, pp 33–41, 2017.

[3] Görlitz Peggy, Grüne Claudia, The Machinery Equipment Industry in Germany, Germany Trade Invest (GTAI), pp 1–16, 2018.

[4] Moyseowicz Andrew, 2016 Top Markets Report Manufacturing Technology, International Trade Administration, pp 1–55, 2016.

[5] Medianas P. Y.,Colombianas E., Carlos J.,Lora T., and Iglesias Pinedo W., Growth determinants in the micro, small and medium size companies in Colombia: the case of the metallurgical sector,Journal Semestre EconÃ³mico, vol. 15, no. 32, pp. 41–76, 2012.

[6] Wan Min, Ye Xiang-Yu, Yang Yun, Zhang Wei-Hong,Theoretical prediction of machining-induced residual stresses in three-dimensional oblique milling processes, International Journal of Mechanical Sciences, vol 133, pp 426–437, 2017.

[7] Sun Yunwen,Guo Qiang, Numerical simulation and prediction of cutting forces in five-axis milling processes with cutter run-out, International Journal of Machine Tools Manufacture, vol 51, pp) 806–815, 2011.

[8] Albertelli Paolo, Keshari Anupam, Matta Andrea, Energy oriented multi cutting parameter optimization in face milling, Journal of Cleaner Production, vol 137, pp 1602e1618, 2016.

[9] Shin Seung-Jun, Woo Jungyub, Rachuri Sudarsan, Energy efficiency of milling machining: Component modeling and online optimization of cutting parameters, Journal of Cleaner Production, vol 161 pp 12–29, 2017.

[10] Liu Dawei, Wang Wei, Wang Lihui, Energy-Efficient Cutting Parameters Determination for NC Machining with Specified Machining Accuracy, Procedia CIRP, Vol 61, pp 523–528, 2017.

[11] Urbikain Gorka, Artetxe Egoitz, López de Lacalle Luis Norberto, Numerical simulation of milling forces with barrel-shaped tools considering runout and tool inclination angles, Applied Mathematical Modelling, vol 47, pp 619–636, 2017.

[12] Ozkirimli Omer, Tunc Lutfi Taner, Budak Erhan, Generalized model for dynamics and stability of multi-axis millingwith complex tool geometries, Journal of Materials Processing Technology, vol 238, pp 446–458, 2016.

[13] Woo Wan-Sik, Lee Choon-Man, A study of the machining characteristics of AISI 1045 steel and Inconel 718 with a cylindrical shape in laser-assisted milling, Applied Thermal Engineering, vol 91, pp 33–42, 2015.

[14] Campocasso Sébastien, Costes Jean-Philippe, Fromentin Guillaume, Bissey-Breton Stéphanie, Poulachon Gérard, Improvement of cutting forces modeling based on oriented cutting tests, Procedia CIRP vol 8, pp 206–211, 2013.

[15] Ginjaume Pujadas Albert, Torre Crespo Felipe, Execution of machining, forming and assembly processes, Editorial: Paraninfo, pages 560, 2005.

[16] Li H., Chen Y. (2014) Machining Process Monitoring. In: Nee A. (eds) Handbook of Manufacturing Engineering and Technology. Springer, London.

[17] Soren T.R., Kumar R., Panigrahi I., Sahoo A.K., Panda A., Das R.K., Machinability behavior of Aluminium Alloys: A Brief Study, Journal Materials Today: Proceedings, vol. 18, N° 7, pp. 5069-5075, 2019

[18] Staroveški T., Brezak D., and Udiljak T., LinuxCNC – The Enhanced Machine Controller: Application and an Overview,Tehnički vjesnik, Vol 20, no 6, pp 1103–1110, 2013.

[19] Huo F., Hong G., and Poo A., "Extended Development of LinuxCNC for Control of a Delta Robot," pp. 22–24, 2015.

[20] Rochler M., "Integration of real-time Ethernet in LinuxCNC Using the example of Sercos III," pp. 1837–1846, 2015.

[21] Ros J., Yoldi R., Plaza A. and Iriarte X., "of a Hexaglide Type Parallel Manipulator on a Real Machine Controller," pp. 587–597.

[22] Fraile J.C., Pérez-Turiel J., González-Sánchez J.L., López-Cruzado J., Rodríguez J.L., Experiences in the development of a robotic application with force control for bone drilling, Journal Iberoamericana de Automática e Informática Industrial RIAI,Volume 5, no 2,pp 93–106, 2008.

[23] Hai-peng H., Guan-xin C., Zhen-long W., "Development of a CNC System for Multi-Axis EDM Based on RT-Linux System Based on RT-Linux, World Congress on Software Engineering, vol 3, pp 211–216, 2009.

# Chapter 5

# Modeling Lost-Circulation into Fractured Formation in Rock Drilling Operations

*Rami Albattat and Hussein Hoteit*

## Abstract

Loss of circulation while drilling is a challenging problem that may interrupt drilling operations, reduce efficiency, and increases cost. When a drilled borehole intercepts conductive faults or fractures, lost circulation manifests as a partial or total escape of drilling, workover, or cementing fluids into the surrounding rock formations. Studying drilling fluid loss into a fractured system has been investigated using laboratory experiments, analytical modeling, and numerical simulations. Analytical modeling of fluid flow is a tool that can be quickly deployed to assess lost circulation and perform diagnostics, including leakage rate decline and fracture conductivity. In this chapter, various analytical methods developed to model the flow of non-Newtonian drilling fluid in a fractured medium are discussed. The solution methods are applicable for yield-power-law, including shear-thinning, shear-thickening, and Bingham plastic fluids. Numerical solutions of the Cauchy equation are used to verify the analytical solutions. Type-curves are also described using dimensionless groups. The solution methods are used to estimate the range of fracture conductivity and time-dependent fluid loss rate, and the ultimate total volume of lost fluid. The applicability of the proposed models is demonstrated for several field cases encountering lost circulations.

**Keywords:** lost-circulation, mud loss, leakage, fractures, Herschel-Bulkley, yield-power law, Cauchy momentum equation, type-curves, non-Newtonian fluids

## 1. Introduction

Drilling technology has been widely deployed in many industries, such as oil and gas, geothermal, environmental remediation, mining, carbon dioxide sequestration, gas storage, water well, infrastructure development, among others [1]. In many situations, drilling entails various technical challenges and difficulties, often causing economic, safety, and environmental disturbances. Due to the complex nature of subsurface geological formations, technical problems often emerge unexpectedly. One of the most pressing problems is lost circulation into fractured formation. Preserving the drilling fluid within the borehole is crucial for removing cuttings, lubrication, hydraulic rotations, pressure control, among others. Fluid total or partial loss into the wellbore surrounding formation may result in wellbore instability. Drilling fluid-loss is a costly problem. This phenomenon may obstruct operations, increase the nonproductive time (NPT), contaminate water tables, and cause formation damage and safety hazard [2–9]. Drilling fluid typically accounts for

25–40% of the total drilling costs [8]. Furthermore, lost circulation may cause other issues, such as wellbore instability, sloughing shale, and washout [10]. Such problems cost the industry about one billion US dollars per annum worldwide [11, 12]. For instance, a study including 1500 gas wells in the Gulf of Mexico showed that lost-circulation accounts for 13% of the total drilling problems [5], as appears in **Figure 1**. Another study for 103 wells in the Duvernay area in Canada reported a loss of $2.6 million and 27.5 days of NPT because of lost circulation (**Figure 2**). In the Middle East, a group of 144 wells in Rumaila field, Iraq, encountered major lost-circulation problems [14], causing 48% of all drilling issues and loss of 295 days NPT, as illustrated in **Figure 3**. This high occurrence of lost-circulation was attributed to the presence of conductive natural fractures in the carbonate formations. For instance, more than 35% of drilled wells in a fractured carbonate formation in Iran experienced lost circulation [15]. Similarly, in Saudi Arabia, one-third of the

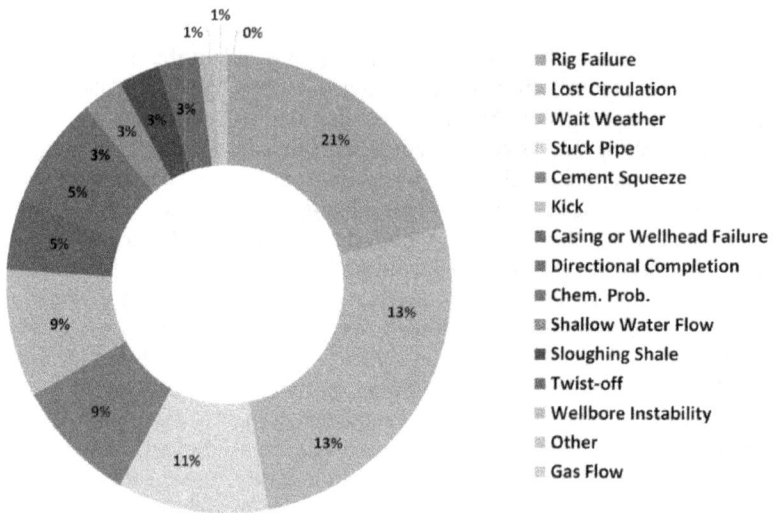

**Figure 1.**
*Contribution of various drilling problems reported for 1500 wells in the Gulf of Mexico [5], where lost circulation accounts for 13% of the total.*

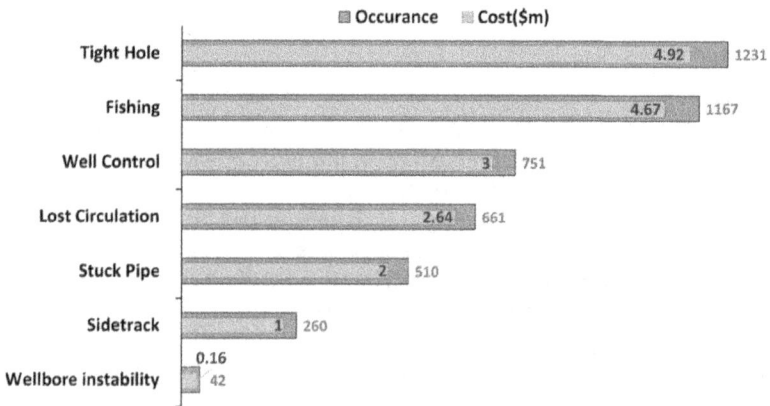

**Figure 2.**
*Cost overhead from various drilling problems for 103 wells in the Duvernay area in Canada [13], where lost circulation account for $2.6 m.*

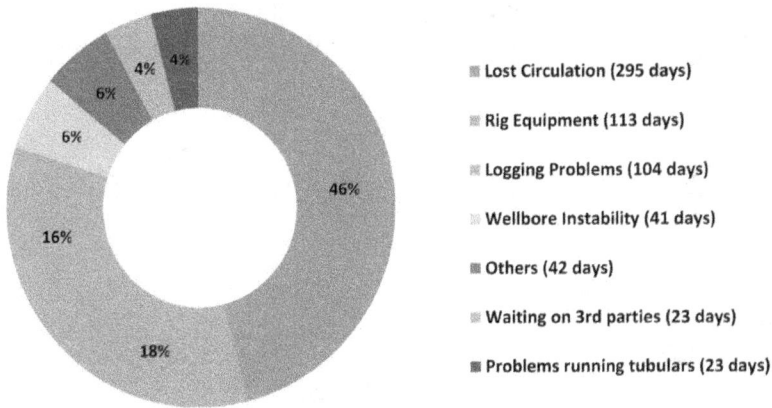

- Lost Circulation (295 days)
- Rig Equipment (113 days)
- Logging Problems (104 days)
- Wellbore Instability (41 days)
- Others (42 days)
- Waiting on 3rd parties (23 days)
- Problems running tubulars (23 days)

**Figure 3.**
*Contribution of various drilling problems to NPT in the Rumaila field in Iraq, total lost circulation is the top drilling issue encountered in that field, data source from [14].*

drilled well underwent lost circulation [16]. Lost circulation has also been reported in other places [17, 18].

Lost circulation can be classified based on the severity of the flow loss rate [19], as shown in **Table 1**. These categories provide general guidance to set the mitigation plan depending on the severity of the fluid loss. Seepage loss can sometimes be tolerable to continue drilling without interruption. However, major loss demands a careful response to regain a full circulation fluid.

Fluid loss is often encountered in fractured formations, which can be subdivided into four types: induced-drilling fracture, caves or vugs, natural fracture, and high permeable formation [20]. An illustration of the four types is shown in **Figure 4**. Each formation type exhibits different fluid loss behavior and, therefore, may require a specific mitigation plan.

Naturally fractured formation, which is the focus of this chapter, is often prone to severe loss during either drilling, cementing, or completion/workover job [21]. Consequently, multiple problems may emanate because of the loss severity, such as kicks, wellbore instability, environmental contamination, and formation damage. To mitigate this problem, one procedure is to add lost circulation material (LCM) to the circulation fluid [22–25]. The LCM fluid properties such as viscosity and density are selected according to the formation type and the subsurface conditions, such as the depth, pressure, and fracture conductivity, among other factors [26]. Fluid loss rate into natural fractures mostly commences with a sudden spurt followed by a gradual declining loss [27]. The ultimately lost volume is dependent on several factors, such as fluid mobility, fracture conductivity, pore-volume, and fracture extension [28].

Because of the nature of this problem that requires immediate intervention, there is a need in the industry to establish an accurate and efficient modeling tool

| Lost Type | Flow Rate Amount |
|-----------|------------------|
| Seepage loss | Less than 1.5 m³/hr |
| Partial loss | 1.5 to 15 m³/hr |
| Severe loss | 15 to 75 m³/hr |
| Total loss | No return to surface |

**Table 1.**
*Classification of lost-circulation severity [19].*

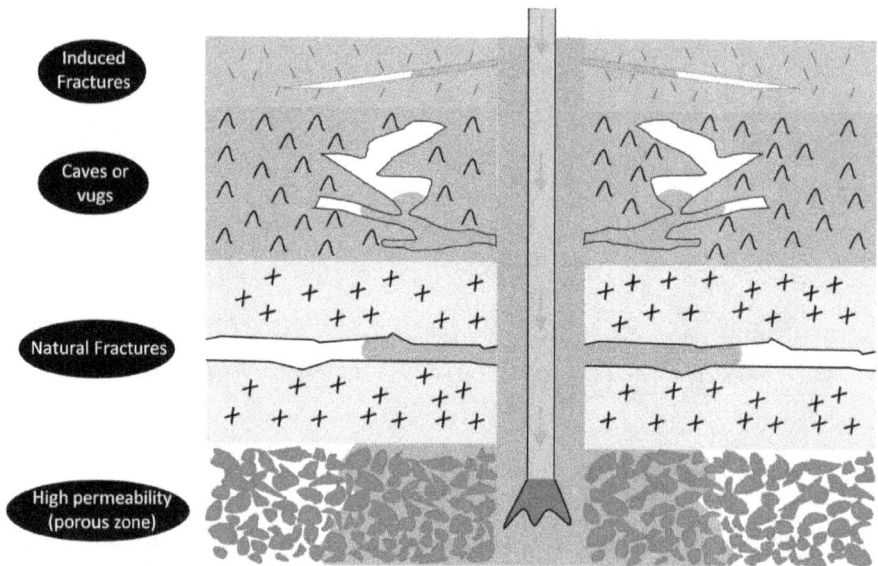

**Figure 4.**
*Four types of rock formations causing loss-circulation. The arrows indicate the direction of circulating fluid starting from the surface within the drill pipe going to the open-hole and to the annulus of the wellbore back to the surface.*

that is feasible at real-time drilling operations to perform diagnostics and predictions.

During the last two decades, many analytical solutions have been introduced in the literature to model mud loss into fractured formation. Early modeling attempts for simplified fractured cases were based on Darcy's Law at steady-state conditions [29, 30]. Afterward, a semi-analytical solution was introduced for the Newtonian fluid model into a horizontal fracture by combining the diffusivity equation and mass conservation in a one-dimensional (1D) radial system [31]. Since this derived ordinary differential equation (ODE) was solved numerically, an analytical solution of the diffusivity equation for a fluid with a constant viscosity at steady-state conditions was introduced [32]. Another approach based on type-curves, which were generated by numerical solutions to describe mud loss volume as a function of time into a horizontal fracture, was established [33]. The type-curves are applicable for non-Newtonian fluid and follow a model that exhibits Bingham plastic rheological behavior. The numerically generated type-curves are based on dimensionless parameters that depend on the effective fracture hydraulic aperture, fluid properties, and differential pressure. However, these models inherit the limitations of numerical methods in introducing numerical artifacts such as numerical dispersion and grid dependency. Later, an analytical solution was proposed [34, 35]. Estimation of hydraulic fracture aperture by simplifying insignificant terms in the final equation form was analyzed [36]. Following the same workflow, a solution was developed based on Yield-Power-Law fluid by reducing a Taylor expansion of the governing nonlinear flow equation into its linear terms [37].

This chapter is intended to give an overview of laboratory and modeling tools applicable for lost circulation in fractured media. Two main subjects are discussed. First, the governing equations to model non-Newtonian fluid in a fractured system are reviewed. We then introduce the solution method to develop a semi-analytical solution to model drilling fluid loss. The fluid exhibits a Herschel-Bulkley behavior, where the Cauchy equation of motion is used to describe the fluid flow. Due to the nonlinearity of the problem, the system of equations is reformulated and transformed into ODE's, which is then computed numerically with an efficient ODE solver [38]. Based on the semi-analytical solution, type-curves are generated, capturing dimensionless fluid loss volume as a function of time. High-resolution finite element methods are used to verify the analytical approach. The applicability of the method is then demonstrated for field cases exhibiting loss of circulation, where formation and fluid uncertainties are addressed with Monte Carlo simulations. In the second subject, an experimental study designed to mimic fluid leakage in a horizontal fracture is discussed. These experiments are used to study the steady-state flow conditions of non-Newtonian fluids into the fracture and demonstrate the flow stoppage process. Simulations are used to replicate the physics, including the effect of fracture deformation. Type-curves are also derived from Cauchy equation of motion to capture the effect of fracture ballooning.

## 2. Mud invasion into a fractured system

Various studies have been conducted in the literature to investigate the flow behavior of drilling fluids in fractured systems [39–41], where a horizontal, radial fracture is considered, as shown in **Figure 5**. The choice of the fracture geometry is motivated by its convenience to be replicated with experimental apparatus and analytical modeling. Even though the fracture geometry may seem simplistic, it can provide useful insights at lab and field scales [43, 44].

Consider two parallel radial plates to mimic a horizontal fracture, intercepting a wellbore, as illustrated in **Figure 5**. Note that horizontal fractures could occur at shallow depths and over-pressurized formations [45, 46].

The general governing equation describing the dynamics of non-Newtonian fluid flow in an open fracture is given by the Cauchy momentum Equation [47, 48], such that,

$$\rho \frac{\partial \mathbf{v}}{\partial t} + \rho(\mathbf{v} \cdot \nabla)\mathbf{v} = \nabla \cdot (-p\mathbf{I} + \boldsymbol{\tau}) + \rho \mathbf{g}, \tag{1}$$

Where $p$ denotes the fluid pressure, t is time, $\boldsymbol{\tau}$ is the shear stress, $\mathbf{g}$ is the gravity term, $\rho$ is the density, $\mathbf{I}$ is identity matrix, and $\mathbf{v}$ is the flow velocity. Divergence delta operator is $\nabla \cdot$.

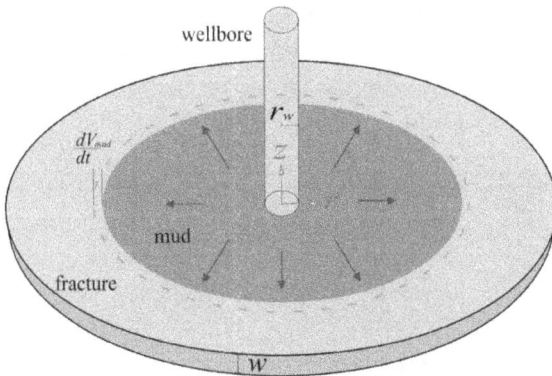

**Figure 5.**
*Physical domain illustrating a horizontal fracture intercepted by a wellbore. The shaded brown area reflects the invaded mud zone, $r_w$ is the wellbore radius, w is the fracture aperture, $V_{mud}$ is the mud loss volume, and t is the time [42].*

Assuming steady-state conditions and neglecting the gravity and inertial effects, Eq. (1) simplifies to,

$$0 = \nabla \cdot (-p\mathbf{I} + \tau), \qquad (2)$$

The above equation reduces into two fundamental forces; pressure forces and shear stress force. In a 1D radial system with polar coordinates, Eq. (2) simplifies to (see [49]),

$$\tau(z, r) = z \frac{\partial p}{\partial r}, \qquad (3)$$

Where $\tau(z, r)$ is the radial shear stress component perpendicular to the $z$-directionand $r$ is radial distance which is the variable argument.

On the other hand, the Herschel-Bulkley model is expressed by [50],

$$\tau(z, r) = \tau_0 + m \left( \frac{dv_r}{dz} \right)^n. \qquad (4)$$

The fluid yield stress, which determines the fluidity state, is denoted by $\tau_0$, and the consistency multiplier and behavioral flow index are $m$and $n$, respectively. The flow index is a positive number, reflecting the fluid rheological behavior where shear-thinning ($n < 1$) and shear-thickening ($n > 1$) can occur. Typical values of this dimensionless parameter for drilling fluids range from 0.3 to 1.0 [51]. Shear rate, which is the derivative of the radial velocity $v_r$ in the $z$-direction, is nonlinear due to the flow index.

The mechanisms corresponding to the solution method of the mud invasion phenomenon in a fractured system are shown on a cross-section in **Figure 6**. The radial velocity decreases as the fluid propagates away from the wellbore within the fracture, and therefore shear stress lessens. Therefore, shear-thinning is expected to be maximum near the wellbore and reduces gradually with radial distance, which induces shear-thickening from yield stress. Furthermore, flow velocity and shear stress variations in the z-direction create layers of fluid rheological properties and fluid self-friction that becomes maximum at the walls of the fracture. Fluid self-friction is minimum at the centerline of the fracture, as shown in **Figure 6**, resulting in a region at the fracture center with zero shear rate, that is, $dv_r/dz = 0$. In this zone, the yield shear stress corresponds to $\tau_0$ (see Eq. (4)). This fluid flow region in the fracture is subdivided into plug-flow and free-flow regions. The plug-region extends toward the walls of the fracture as the shear-stress reduces, and the fluid flows further from the wellbore. The plug region can eventually reach the fracture wall leading to a complete stoppage of the fluid leakage (see **Figure 7a**). In other words,

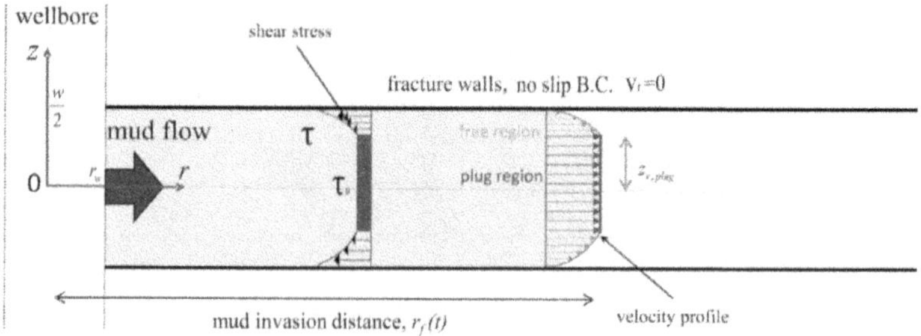

**Figure 6.**
*Illustration of an infinite-acting fracture of aperture w, intercepting a wellbore with radius $r_w$. No-flow and no-slip boundary conditions are imposed at the fracture wall. Mud-front distance $r_f(t)$ is a time-dependent parameter.*

mud leakage stalls in the fracture when the pressure gradient between the fracture inlet and the mud front becomes smaller than the yield stress $\tau_0$ (see **Figure 7b**).

The following boundary conditions are considered for the two flow regions,

$$
v_r(z) = \begin{cases} v_{r,plug}(z), & for \quad z \leq z_{plug} \\ v_{r,free}(z), & for \quad z_{plug} < z < \dfrac{w}{2} \\ 0, & for \quad z = \dfrac{w}{2} \end{cases} \tag{5}
$$

In the above equation, $z_{plug}$ represents the extension of the plug region in the z-direction. $v_{r,plug}$, and $v_{r,free}$ are, respectively, the flow velocities within the plug- and free-region. The no-slip boundary condition is described by the last equation in (5).

Eq. (4) and Eq. (3) are combined to express the solution of the velocity, such that,

$$
v_r(z) = \frac{n\left(-\frac{\partial p}{\partial r}\frac{w}{2} + \tau_0\right)\left(\frac{\frac{\partial p}{\partial z}\frac{w}{2}-\tau_0}{m}\right)^{1/n} + n\left(-\frac{\partial p}{\partial r}z + \tau_0\right)\left(\frac{\frac{\partial p}{\partial z}z-\tau_0}{m}\right)^{\frac{1}{n}}}{\frac{\partial p}{\partial r}(n+1)} \tag{6}
$$

The plug-region is modeled by imposing the condition, $dv_r/dz = 0$. Therefore, Eq. (6) can be expressed for each region individually, as follows,

$$
v_{r,free}(z) = \frac{n}{n+1}\left(z_{plug} - \frac{w}{2}\right)\left(\frac{\frac{\partial p}{\partial r}\left(\frac{w}{2}-z_{plug}\right)}{m}\right)^{1/n} + \frac{n}{n+1}(z - z_{plug})\left(\frac{\frac{\partial p}{\partial r}(z - z_{plug})}{m}\right)^{\frac{1}{n}}
$$

$$
v_{r,plug}(z) = \frac{n}{n+1}\left(\frac{\tau_0}{\frac{\partial p}{\partial r}} - \frac{w}{2}\right)\left(\frac{\left(\frac{w}{2}\frac{\partial p}{\partial r}-\tau_0\right)}{m}\right)^{1/n}
$$

$$\tag{7}$$

From the definition of the total volumetric flow rate $Q_{total}$, one obtains,

$$
Q_{total} = Q_{plug} + Q_{free} \tag{8}
$$

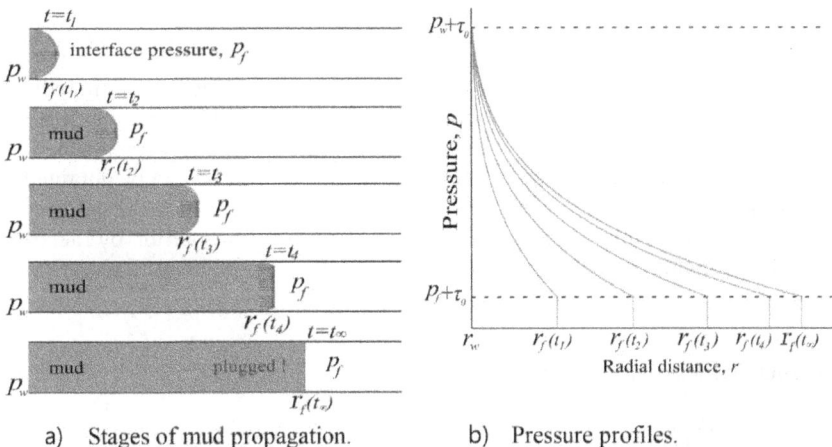

a) Stages of mud propagation.

b) Pressure profiles.

**Figure 7.**
*Illustration of yield-power-law fluid flow in a radial fracture showing the evolution of the propagation of the plug region as the mud travels away from the wellbore, resulting in total plugging (a). Plot (b) shows typical pressure profiles versus radial distance at various times and invasion distances.*

Applying surface integral,

$$Q_{total} = 4\pi r \int_0^{z_{plug}} v_{r,plug} dz + 4\pi r \int_{z_{plug}}^{w/2} v_{r,free} dz \tag{9}$$

Substituting Eq. (7) into Eq. (9) and arranging to obtain,

$$Q_{total}^n = \frac{(4\pi r)^n}{m} \left(\frac{w}{2}\right)^{2n+1} \left(\frac{n}{2n+1}\right)^n \left(\frac{dp}{dr}\right)^n \left(1 - \frac{\tau_0}{\frac{w}{2}\frac{dp}{dr}}\right) \left(1 - \left(\frac{1}{n+1}\right)\frac{\tau_0}{\frac{w}{2}\frac{dp}{dr}} - \left(\frac{n}{n+1}\right)\left(\frac{\tau_0}{\frac{w}{2}\frac{dp}{dr}}\right)^2\right)^n \tag{10}$$

The above equation is rearranged with the following quadratic equation to express the pressure term explicitly, that is,

$$\left(\frac{dp}{dr}\right)^2 - \left(\frac{Q_{total}^n}{r^n A} + B\right)\frac{dp}{dr} + D = 0 \tag{11}$$

Where,

$$A = \frac{(4\pi)^n}{m}\left(\frac{w}{2}\right)^{2n+1}\left(\frac{n}{2n+1}\right)^n; B = \left(\frac{2n+1}{n+1}\right)\frac{\tau_0}{w/2}; D = \left(\frac{n-n^2}{n+1}\right)\left(\frac{\tau_0}{w/2}\right)^2$$

Solving the differential pressure and integrating along the radial domain by implementing a moving boundary condition, a final ODE system is reached, as follows,

$$\begin{cases} p_f - p_w = \dfrac{B\left(r_f(t) - r_w\right)}{2} + \dfrac{Q_{total}^n\left(r_f(t)^{1-n} - r_w^{1-n}\right)}{2(1-n)A} + \dfrac{1}{2}\int_{r_w}^{r_f(t)}\left(\sqrt{\left(B + \dfrac{Q_{total}^n}{r^n A}\right)^2 - 4D}\right)dr \\ \\ Q_{total} = 2\pi w r_f(t)\dfrac{dr_f(t)}{dt} \end{cases} \tag{12}$$

Eq. (12) is nonlinear for a general value of $n$, and it cannot be solved analytically. However, a general semi-analytical solution can be derived [42]. This solution is a generalization to other particular solutions in the literature. For instance, when $n = 1$, reflecting a Bingham plastic fluid, a closed-form solution can be obtained, as demonstrated by Lietard *et al.* [33]. **Figure 8** shows that the proposed general semi-analytical solution is in perfect agreement with the analytical solution by Lietard *et al.* [33]. For general cases of $n$, numerical simulations could be used to verify the semi-analytical model, as shown in **Figure 9**.

### 2.1 Dimensionless type-curves

For general applications, type-curves are used as a diagnostic tool to assess the solution by matching the trends of observed data to the type-curves. This approach is commonly used in well testing [52]. To enable scalability of the solution for a wide range of problem conditions, type-curves are expressed in terms of dimensionless groups. In this problem, the following dimensionless variables are considered,

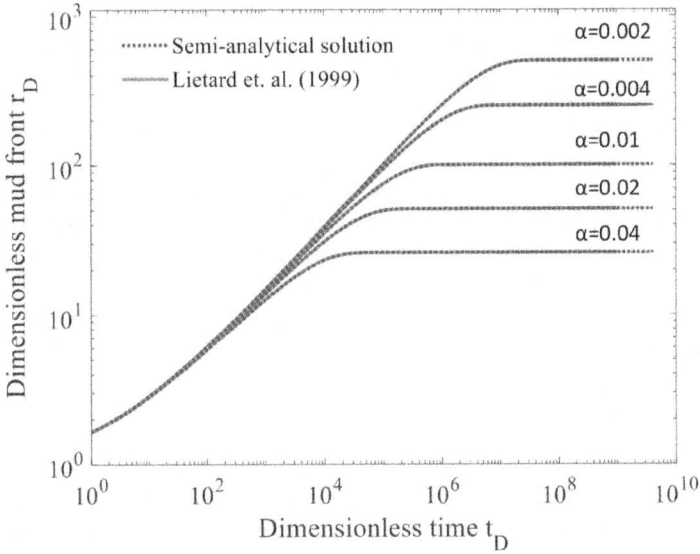

**Figure 8.**
*Proposed semi-analytical solution compared with Liétard et al. (1999) for a Bingham plastic fluid [42].*

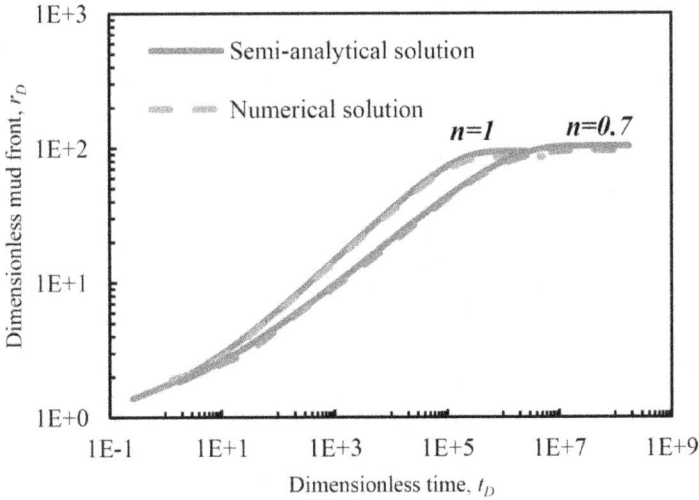

**Figure 9.**
*Proposed semi-analytical solution compared with a finite-element model showing a good agreement [42].*

$$r_D = \frac{r_f}{r_w}$$

$$V_D = \frac{V_m}{V_w} = \frac{\pi w \left(r_f{}^2 - r_w{}^2\right)}{\pi w r_w{}^2} = \left(\frac{r_f}{r_w}\right)^2 - 1 = r_D{}^2 - 1$$

$$\alpha = \left(\frac{2n+1}{n+1}\right)\left(\frac{2r_w}{w}\right)\left(\frac{\tau_0}{\Delta p}\right) \tag{13}$$

$$\beta = \left(\frac{n}{2n+1}\right)\left(\frac{w}{r_w}\right)^{1+\frac{1}{n}}\left(\frac{\Delta p}{m}\right)^{\frac{1}{n}}$$

$$t_D = t\beta$$

Where, $r_D$ denotes the dimensionless radial mud-front, $V_D$ is the dimensionless mud-loss volume, and $t_D$ is the dimensionless time. The generated type-curves from Eq. (13) can represent a range of fluid properties, captured by two parameters. The first parameter, $\alpha$, represents the mud rheological properties and flow behavior, while the second parameter, $\beta$, reflects the criteria for mud loss stoppage. The final forms of the type-curves are illustrated in **Figure 10**. These type-curves for mud-loss are more accurate than the ones proposed by Majidi et al. [37], which showed lower accuracy for small $\alpha$. Comparisons of the proposed solution with Majidi et al. is plotted in **Figure 11**, which show that the simplified model by Majidi et al. could overestimate the radial distance of mud invasion.

## 2.2 Demonstration for real field cases

The applicability of the discussed modeling approach is demonstrated for field cases that encountered lost circulation. The field data for four wells include the leakage rates and the fluid types corresponding to Bingham plastic and Herschel-Bulkley fluids. Two modeling approaches are discussed, where the first is a deterministic approach that is matched with the proposed analytical solution, and the second is a probabilistic approach based on Monte Carlo simulations. The utilization of the probabilistic approach is motivated by the subsurface uncertainty of the downhole parameters.

### 2.2.1 Field case 1

The lost circulation data correspond to two wells, Machar 18 and Machar 20 in Machar field in North Sea [33]. The fluid loss occurred due to the presence of natural fractures. Using the proposed type-curves, the fracture apertures were

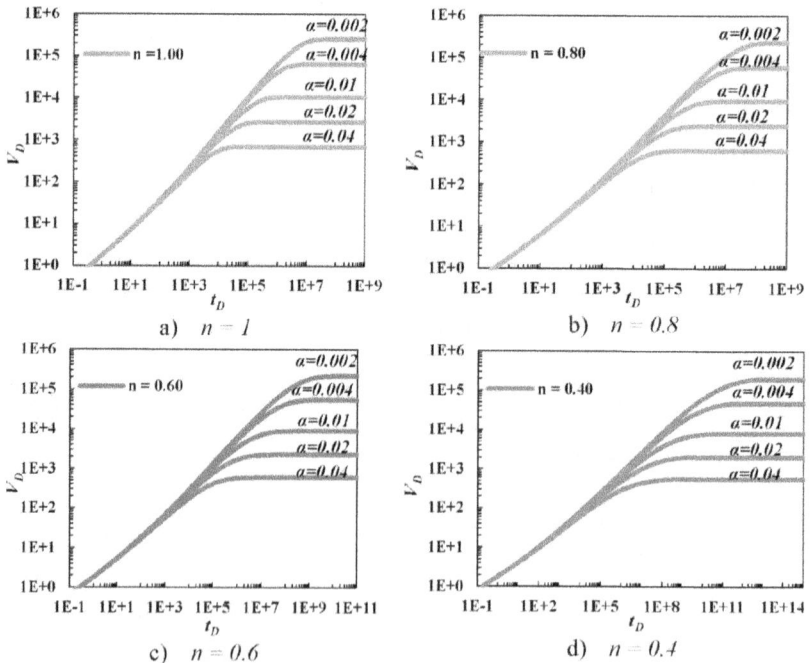

**Figure 10.**
*Dimensionless type-curves showing dimensionless mud loss volume versus dimensionless elapsed time for a set of n and α [42].*

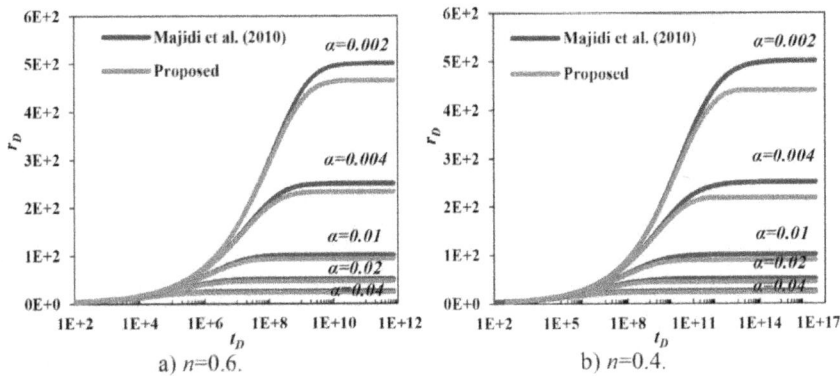

a) $n=0.6$.  b) $n=0.4$.

**Figure 11.**
*Comparison between the proposed solution and the one by Majidi et al., which tends to overestimate the invasion distance for low values of $\alpha$ [42].*

determined to be 0.42 and 0.61 mm, respectively. Both wells are plotted with the semi-analytical trends **Figure 12**. The relevant fluid and formation properties are given in **Table 2**.

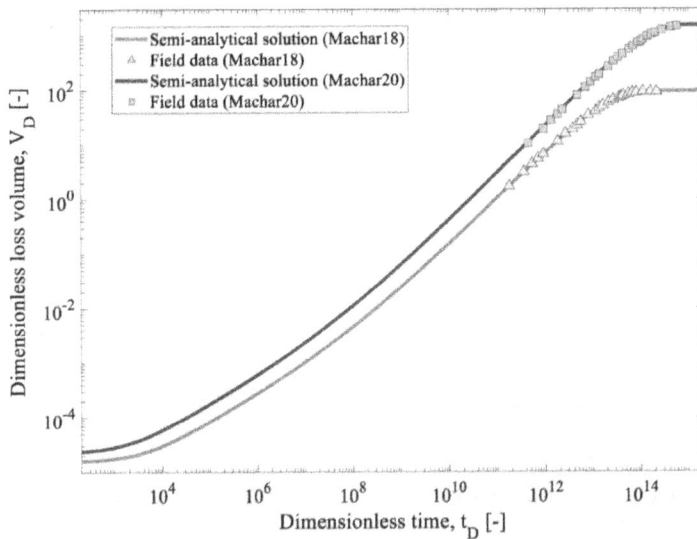

**Figure 12.**
*Real field data matched with the semi-analytical solutions for two wells; well data are from Lietard et al. [33].*

| Property | Machar 18 | Machar 20 |
|---|---|---|
| Dimensionless parameter $\alpha$ | 0.00215 | 0.0006436 |
| Fracture aperture $w$ [mm] | 0.425 | 0.616 |
| Fracture aperture $w$ [mm] | 0.42 | 0.64 |
| Flow behavioral index | 1 | |
| Fluid yield stress [Pa] | 9.34 | |
| Fluid plastic viscosity [Pa.s] | 0.035 | |

**Table 2.**
*Fluid properties and fracture aperture for the two wells.*

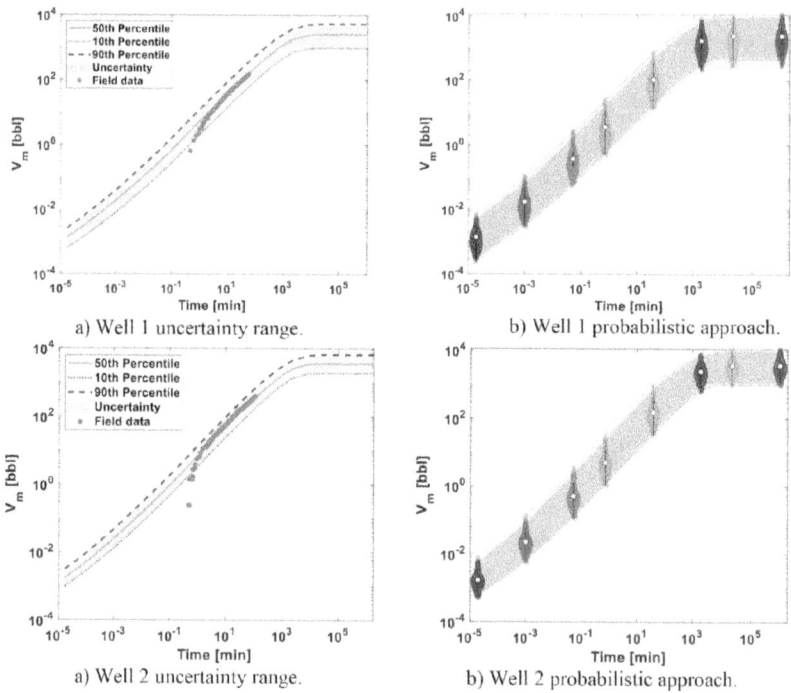

a) Well 1 uncertainty range.

b) Well 1 probabilistic approach.

a) Well 2 uncertainty range.

b) Well 2 probabilistic approach.

**Figure 13.**
*Both wells are plotted cumulative mud loss volume versus time showing uncertainty plots (a) and (c), and violin distributions at selected time points (b) and (d) for well 1 and 2 respectively [42].*

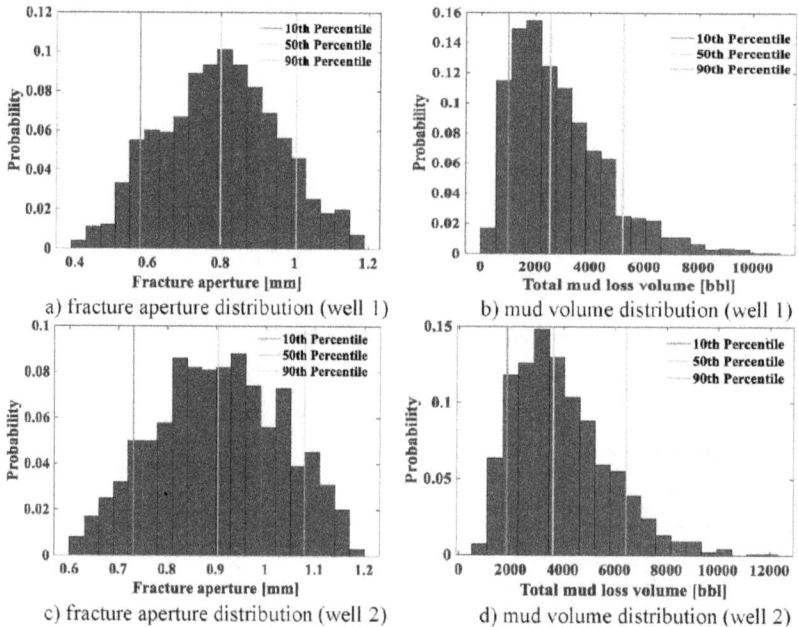

a) fracture aperture distribution (well 1)

b) mud volume distribution (well 1)

c) fracture aperture distribution (well 2)

d) mud volume distribution (well 2)

**Figure 14.**
*Well 1 ((a) and (b)) and well 2 ((c) and (d)) output results showing average hydraulic fracture aperture in mm can be obtained from the mud-loss analysis plus maximum cumulative mud loss volume [42].*

| Property | Well 1&2 |
|---|---|
| Pressure difference [MPa] | 4.83–5.51 |
| Flow behavioral index | 0.94 |
| Consistency [Pa.s$^n$] | 3.83 |
| Fluid yield stress [Pa] | 4.02 |

**Table 3.**
*Wells condition and fluid properties for field case 2 at surface conditions.*

### 2.2.2 Field case 2

In this case, two sets of data were reported for two wells in the Gulf of Mexico [53]. The drilling fluid is Herschel-Bulkley fluid. The leakage data is provided as a volumetric flow rate in gallon per minute versus time. The provided data is converted into volumetric flow rates in m$^3$ per second as a function of time. Due to the uncertainty of subsurface parameters such as pressure drop, and rock properties, the semi-analytical solution is combined Monte Carlo simulations. The solutions and the data match are shown in **Figure 13**. Various useful information can be generated from these simulations. For instance, the uncertainty range of the hydraulic fracture aperture for the natural fractures can be predicted. Furthermore, the total mud-loss volumes can be estimated in the form of a histogram covering the range of uncertainty for P10, P50, and P90, percentiles, as shown in **Figure 14**. The relevant data are provided in **Table 3**.

## 3. Experimental study

The flow of drilling mud into an artificial fracture was studied by Majidi et al. [54]. The fracture consisted of two horizontal plates placed in parallel within a 1 mm opening, mimicking a fracture aperture. An illustration and photograph of the experimental apparatus are shown in **Figure 15**. Different fluid types and flow conditions were investigated. The inner hole, used for fluid injection, is 3 inches (in) in diameter located at the center of the upper plate. The radius of the fracture plates is 36 in. Pressure transducers were installed at different locations along the fracture radial distance. The pressure at the fracture outlet was maintained at the

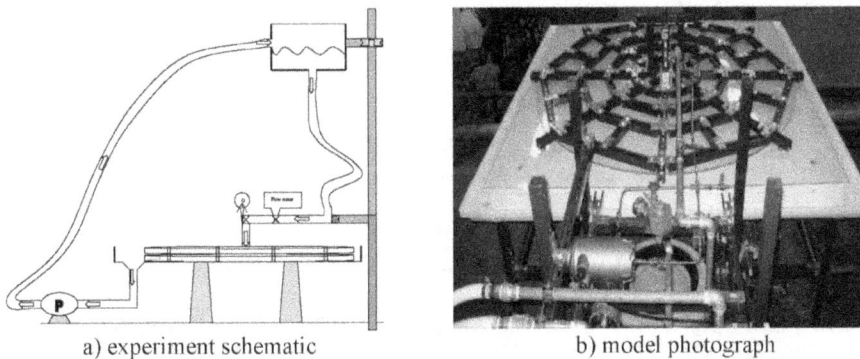

a) experiment schematic          b) model photograph

**Figure 15.**
*Experimental setup showing a schematic view (a) and a photograph (b) of an experimental setup consisting of two parallel plates separated by a gap [54] (with permission from the author).*

atmospheric conditions. The non-Newtonian fluid was injected by gravity from a supply tank placed at a certain height.

## 3.1 Steady-state flow test

Fluids were injected into the fracture system at a constant rate in a circulation mode until the steady-state condition was reached for the pressure within the fracture. The experiment was conducted with two different fluids, as provided in **Table 4**. A semi-analytical solution was developed to model steady-state radial flow for non-Newtonian fluids following a Yield-Power Law model described in the following,

$$-\frac{dp}{dr} = \frac{kQ_{in}^n}{\left[4\pi\left(\frac{n}{2n+1}\right)\left(\frac{w}{2}\right)^2\right]^n\left(\frac{w}{2}\right)} + \left(\frac{2n+1}{n+1}\right)\left(\frac{2\tau_0}{w}\right) \qquad (14)$$

Where differential pressure with respect to radial distance $\frac{dp}{dr}$ equals to the operating conditions for inlet flow rate $Q_{in}$, fluid properties, and fracture aperture $w$. The fluid is characterized by fluid yield stress $\tau_0$, consistency factor $k$, and flow behavioral index $n$. The limitation of this semi-analytical model is related to the assumption of constant fracture aperture, which is inconsistent with the varying fracture aperture caused by a slight deformation of the upper plate due to fluid pressure during injection.

A commercial simulator (COMSOL®) was used to investigate the effect of fracture wall deformation on the pressure behavior [55]. The results of viscosity and corresponding shear at three locations with respect to radial distance, reflecting shear thinning and thickening effects are shown in **Figure 16**.

## 3.2 Fracture ballooning effect

Fracture deformation and ballooning due to increased fluid pressure inside the fracture could occur [55]. Majidi et al. reported that there is a deformation happening somewhere between inlet and outlet caused by force distribution from fluid pressure. The physics behind fracture ballooning is related to mechanical deformation of the fracture resulting from fluid pressure and the surrounding stress field, causing the fracture to reshape its aperture by either opening or closing [56–58]. Simulations are used to investigate the fracture wall deformation and the corresponding pressure response. Simulations were conducted for deformed fracture plates, as illustrated in **Figure 17**. The results of pressure profiles for the deformed plate are in good agreement with the experimental measurements, as observed in **Figure 18**.

| Fluid Name | Polymer solution (Pack) | Xanthan gum |
| --- | --- | --- |
| Yield-Stress, $\tau_0$ [Pa] | 0 | 1.45 |
| Consistency factor, $k$ [Pa.s$^n$] | 5.30 | 3.37 |
| Flow Index, $n$ [Dimensionless] | 0.480 | 0.407 |
| Flow Rate, $Q_{in}$ [m$^3$/hr] | 0.057, 0.113, 0.237 | 0.025, 0.057, 0.123 |
| Flow Rate, $Q_{in}$ [US gal/min] | 0.25, 0.50, 1.00 | 0.11, 0.25, 0.54 |

**Table 4.**
*Rheological properties for two fluids used in steady-state flow test [54].*

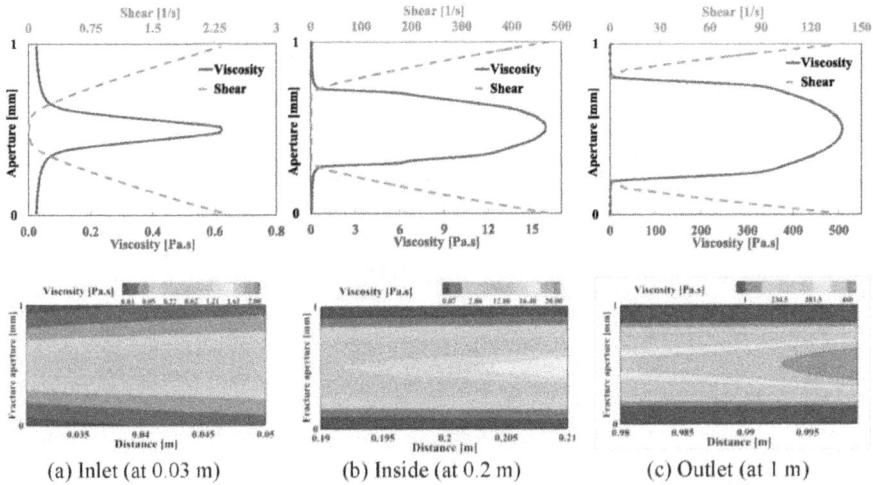

**Figure 16.**
*Fluid viscosity and shear rate along the fracture aperture at the inlet (0.03 m), inside (0.2 m), and outlet (1 m). The bottom plots show the corresponding viscosity color maps from the numerical model.*

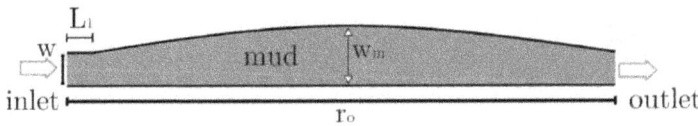

**Figure 17.**
*Cross-section of proposed geometry showing a parabolic-shape deformation for upper plate [55].*

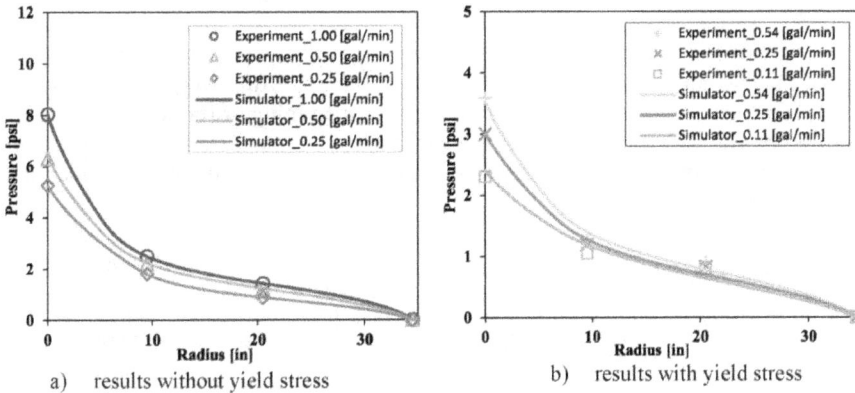

**Figure 18.**
*Match results of experimental data and simulations for the six cases at various flow rates, without fluid yield stress (a) with yield stress (b) [55].*

## 3.3 Plug flow test

The experiment was conducted to investigate the stoppage of flow invasion as a result of the fluid yield stress. The final invasion distances for two fluids were measured in the radial fracture system. Different injection pressures at the inlet were tested, for which the steady-state mud invasion front was measured. This

$$\Delta p_3 > \Delta p_2 > \Delta p_1$$

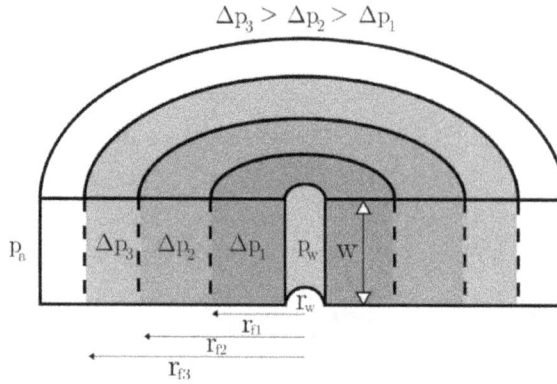

**Figure 19.**
*Illustration of a plug-test experiment showing mud-invasion fronts $r_f$ in different colors at different applied pressure.*

| Fluid Name | Xanthan gum 40 [g/L] | Xanthan gum 30 [g/L] |
|---|---|---|
| Yield-Stress, $\tau_0$ [Pa] | 13.9 | 9.6 |
| Consistency factor, $k$ [Pa.s$^n$] | 4 | 5.5 |
| Flow Index, $n$ [Dimensionless] | 0.34 | 0.315 |
| Parabola vertex, $w_m$ [mm] | 1.5 | 1.25 |

**Table 5.**
*Fluid properties used in plug flow experiment.*

experiment is illustrated in **Figure 19**. Understanding this effect is required to predict the maximum mud loss volume [59]. The two fluids are described in **Table 5**.

Because of the fracture ballooning, the aperture is not constant, as shown in **Figure 18**. The equation used to describe the varying aperture $w(r)$ as a function of radial distance $r$ is given in Eq. (15). The maximum opening of fracture aperture is captured by the parameter $w_m$, distance from inlet where no occurred deformation is defined by $L_1$ and initial aperture where no deformation occurred is $w_i$. Inlet and outlet radii are $r_w$ and $r_o$.

$$w(r) = \begin{cases} w_i & \text{if} \quad r_w \leq r < L_1 \\ (w_m - w_i)\left[1 - \frac{\left(r - r_w - L_1 - \frac{r_o - r_w - L_1}{2}\right)^2}{\left(\frac{r_o - r_w - L_1}{2}\right)^2}\right] + w_i & \text{if} \quad r \geq L_1 + r_w \end{cases} \tag{15}$$

When the mud stops, the flow rate is zero, that is,

$$\frac{dp}{dr} = \frac{2\tau_0(1 + 2m)}{(1 + m)w} \tag{16}$$

Integrating the above equation over the fracture to get,

$$\int_{r_w}^{r_f} p(r)dr = \int_{r_w}^{L_1} p(r)dr + \int_{L_1}^{r_f} p(r)dr \tag{17}$$

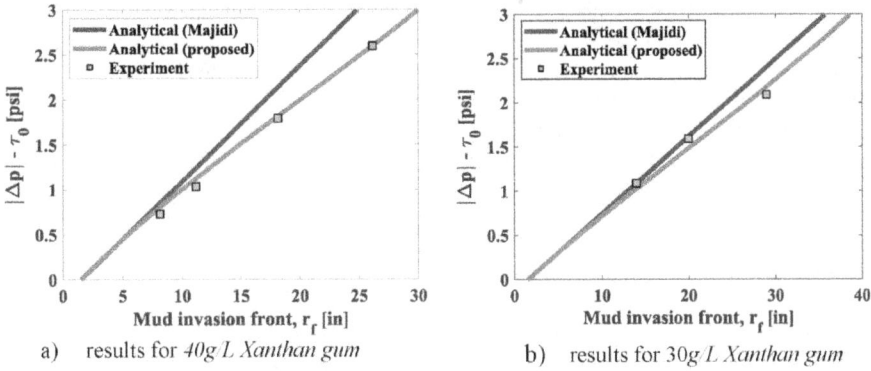

**Figure 20.**
*Mud invasion results for two fluids with yield stress (40 g/L xanthan gum (a); 30 g/L xanthan gum (b)), and comparison with the a different model [55].*

After integration of Eq. (17), the invasion radius, $r_f$, is expressed in terms of the fluid rheological properties and the fracture geometry, as follows,

$$r_f = \frac{1}{2}(L_1 + r_o + r_w) +$$

$$\frac{\frac{1}{2}w_m}{\sqrt{\frac{(w_i - w_m)w_m}{(L_1 - r_o + r_w)^2}}} \tan \left[ \frac{(-2(1+2m)(L_1 - r_w)\tau_0 + \Delta p(1+m)w_i)\sqrt{\frac{(w_i - w_m)w_m}{(L - r_o + r_w)^2}}}{(1+2m)\tau_0 w_i} + \arctan\left[ \frac{(L_1 - r_o - r_w)}{w_m}\sqrt{\frac{(w_i - w_m)w_m}{(L_1 - r_o + r_w)^2}}\right] \right]$$

$$(18)$$

The solution of this equation is plotted along with experimental measurements in **Figure 20**. The results are in excellent agreement with the experimental data and show significant improvement to the model by Majidi et al. [37], which assumes uniform fracture and, therefore, ignores fracture deformation.

## 4. Conclusions

Lost circulation during drilling operations is a common problem that requires immediate intervention to circumvent fluid loss. Diagnostic tools, based on simplified input data such as fluid properties, pressure, and rate trends, can be quickly deployed to quantify uncertainties related to the fluid leakage into the subsurface formation and to perform predictions. Semi-analytical solutions are used to model the leakage behavior of general Herschel-Bulkley fluids into a horizontal infinite-acting fracture, mimicking the effect of a fractured formation. The approach is applicable to different types of non-Newtonian fluids, including yield stress shear-thinning and shear-thickening fluids. The model could predict the trend of mud leakage in a system with horizontal fractures as a function of time. It can estimate the effective hydraulic aperture of the fracture, the ultimate total mud-loss volume, and the expected duration before the leakage stalls, if conditions allow.

Dimensionless groups are used to generate type-curves, which can provide quick diagnostics about the leakage behavior from matching the type-curve trends without a need for simulations. The applicability of the model was demonstrated for four wells from two different fields. A numerical procedure was described to couple the model with Monte-Carlo simulations to perform predictions under uncertainties. This approach is a practical diagnostic tool to perform quick predictions and to optimize LCM selection.

## Acknowledgements

The authors thank King Abdullah University of Science and Technology (KAUST), and Ali I. Al-Naimi Petroleum Engineering Research Center (ANPERC) for supporting this work.

## Conflict of interest

The authors declare no conflict of interest.

## Nomenclature

| | |
|---|---|
| **v** | velocity vector, m/s |
| $A$ | defined constant |
| $B$ | defined constant |
| $D$ | defined constant |
| $d$ | normal derivative |
| **g** | gravitational acceleration, m/s$^2$ |
| **I** | identity matrix |
| $L_1$ | distance from inlet where no occurred deformation along the plate, m |
| $m$ | consistency multiplier, kg/(Pa.s$^n$) |
| $n$ | flow behavioral index |
| $p$ | pressure, psi |
| $p_f$ | formation pressure, psi |
| $p_w$ | wellbore pressure, psi |
| $Q_{free}$ | volumetric flow rate in free region, m$^3$/s |
| $Q_{plug}$ | volumetric flow rate in plug region, m$^3$/s |
| $Q_{total}$ | total volumetric flow rate, m$^3$/s |
| $r$ | radial distance (variable argument), m |
| $r_D$ | dimensionless mud front radius, m |
| $r_f$ | mud front radius, m |
| $r_w$ | wellbore radius, m |
| $t$ | time, min |
| $t_D$ | dimensionless time |
| $V_D$ | dimensionless mud loss volume |
| $V_m$ | cumulative mud loss volume, bbl |
| $V_w$ | wellbore volume due to fracture aperture, bbl |
| $v_r$ | radial velocity, m/s |
| $v_{r,plug}$ | radial velocity in plug region (non-deformed region), m/s |
| $v_{r,free}$ | radial velocity profile in the free deformed region, m/s |
| $w$ | fracture aperture, mm |

| | |
|---|---|
| $w(r)$ | varying aperture as a function of radial distance, mm |
| $w_i$ | initial fracture aperture, mm |
| $w_m$ | maximum opening of fracture aperture, mm |
| $z$ | z-direction in radial coordinate, m |
| $z_{plug}$ | height of plug region profile, m |
| $\alpha$ | dimensionless parameter |
| $\beta$ | dimensionless parameter |
| $\partial$ | partial derivative |
| $\gamma$ | shear rate, 1/s |
| $\Delta p$ | pressure drop, psi |
| $\mu_0$ | viscosity due to fluid yield stress, Pa.s |
| $\mu_{eff}$ | effective viscosity, Pa.s |
| $\rho$ | density |
| $\boldsymbol{\tau}$ | shear stress tensor, Pa |
| $\tau(z,r)$ | shear stress component as a function of z and r, Pa |
| $\tau$ | shear stress, Pa |
| $\tau_0$ | fluid yield stress, Pa |

## Author details

Rami Albattat and Hussein Hoteit*
Physical Science and Engineering Division, King Abdullah University of Science and Technology, Thuwal, Saudi Arabia

*Address all correspondence to: hussein.hoteit@kaust.edu.sa

IntechOpen

# References

[1] N.R. Council. Drilling and excavation technologies for the future. In: Chapter 3. National Academies Press; 1994. pp. 12-29 https://doi.org/https://doi.org/10.17226/2349

[2] J. Amadi-echendu, A.E. Yakubu, Asset Operations: Non-productive Times During Oil Well Drilling, (2016) 43–48. https://doi.org/10.1007/978-3-319-27064-7.

[3] B. Rehm, J. Schubert, A. Haghshenas, A. Paknejad, J. Hughes, Managed Pressure Drilling, Gulf Publ. Co. (2008).

[4] Bolton et al., Dramatic incidents during drilling at Wairakei Geothermal Field, New Zealand, Geothermics. 38 (2009) 40–47.

[5] D. Reid, S. Rosenberg, M. Montgomery, M. Sutherland, P. York, W. Intl, Drilling-Hazard Mitigation—Reducing Nonproductive Time by Application of Common Sense and High-Value-Well Construction Techniques, Offshore Technol. Conf. (2006).

[6] J. Willis, Wynne, Combating Lost Circulation While Cementing in the Mid-Continentt, Am. Pet. Inst. (1959).

[7] R.A. Silent, Circulation Losses, Drill. Prod. Pract. (1936).

[8] E. Lécolier, B. Herzhaft, L. Rousseau, L. Néau, B. Quillien, J. Kieffer, Development of a nanocomposite gel for lost circulation treatment, SPE - Eur. Form. Damage Conf. Proceedings, EFDC. (2005) 327–335. https://doi.org/10.2523/94686-ms.

[9] J. Seyedmohammadi, The effects of drilling fluids and environment protection from pollutants using some models, Model. Earth Syst. Environ. 3 (2017) 23. https://doi.org/10.1007/s40808-017-0299-7.

[10] F.K. Saleh, C. Teodoriu, C.P. Ezeakacha, S. Salehi, Geothermal Drilling : A Review of Drilling Challenges with Mud Design and Lost Circulation Problem, (2020) 1–8.

[11] S. Al Maskary, A. Abdul Halim, S. Al Menhali, Curing losses while drilling & cementing, in: Abu Dhabi Int. Pet. Exhib. Conf., 2014.

[12] N. Droger, K. Eliseeva, L. Todd, C. Ellis, O. Salih, N. Silko, E. Al, Degradable fiber pill for lost circulation in fractured reservoir sections, in: IADC/SPE, 2014.

[13] A. Fox, Can Geomechanics Improve Your Drilling and Completions?, 2018.

[14] U. Arshad, B. Jain, M. Ramzan, W. Alward, L. Diaz, I. Hasan, A. Aliyev, C. Riji, Engineered Solution to Reduce the Impact of Lost Circulation During Drilling and Cementing in Rumaila Field, Iraq, in: International Petroleum Technology Conference, Doha, Qatar, 2015. https://doi.org/10.2523/iptc-18245-ms.

[15] J. Abdollahi, I.M. Carlsen, S. Mjaaland, P. Skalle, A. Rafiei, S. Zarei, Underbalanced drilling as a tool for optimized drilling and completion contingency in fractured carbonate reservoirs, SPE/IADC Underbalanced Technol. Conf. Exhib. - Proc. (2004) 195–204. https://doi.org/10.2523/91579-ms.

[16] M. Ameen, Fracture and in-situ stress patterns and impact on performance in the Khuff structural prospects, eastern offshore Saudi Arabia, Mar Pet. Geol. 50 (2014).

[17] D. Denney, Controlling Lost Circulation in flowing HP/HT Wells: Case History, JPT, J. Pet. Technol. 56 (2004) 55–56. https://doi.org/10.2118/0104-0055-jpt.

[18] Schlumberger, Lost Circulation Solution Saves Time and USD 3.3 Million for Chevron, Schlumberger. (2011).

[19] E.B. Nelson, D. Guillot, Well Cementing: Second Edition, Schlumberger, Texas, 2006.

[20] A. Lavrov, Lost circulation: Mechanisms and solutions, 2016. h ttps://doi.org/10.1016/C2015-0-00926-1.

[21] V. Dokhani, Y. Ma, Z. Li, T. Geng, M. Yu, Transient effects of leak-off and fracture ballooning on mud loss in naturally fractured formations, 53rd U. S. Rock Mech. Symp. (2019).

[22] A. Ali, C.L. Kalloo, U.B. Singh, Preventing lost circulation in severely depleted unoonsolidated sandstone reservoirs, SPE Repr. Ser. (1997) 103–109. doi:10.1016/0148-9062(94) 91150-9.

[23] D.J. Attong, U.B. Singh, G. Teixeira, Successful use of a modified MWD tool in a high-concentration LCM mud system, SPE Drill. Complet. 10 (1995) 22–26. doi:10.2118/25690-PA.

[24] K. Knudsen, G.A. Leon, A.E. Sanabria, A. Ansari, R.M. Pino, First application of thermal activated resin as unconventional LCM in the Middle East, Soc. Pet. Eng. - Abu Dhabi Int. Pet. Exhib. Conf. ADIPEC 2015. (2015) 1–8. doi:10.2118/177430-ms.

[25] M. Olsen, G. Lende, K. Rehman, P. Haugum, J. Mo, G. Smaaskjar, R. Næss, Innovative and established LCM cementing solutions combined to create novel LCM cementing fluid train, Soc. Pet. Eng. - SPE Norw. One Day Semin. 2019. (2019). doi:10.2118/195622-ms.

[26] J. Luzardo, E.P. Oliveira, P.W.J. Derks, R.V. Nascimento, A.P. Gramatges, R. Valle, I.G. Pantano, F. Sbaglia, K. Inderberg, Alternative lost circulation material for depleted reservoirs, in: OTC Bras., Offshore Technology Conference, 2015. doi: 10.4043/26188-MS.

[27] C.G. Dyke, Bailin Wu, D. Milton-Taylor, Advances in characterizing natural-fracture permeability from mud- log data, SPE Form. Eval. 10 (1995) 160–166. doi:10.2118/25022-pa.

[28] J. NORMAN, Coriolis sensors open lines to real-time data, Drill. Contract. 67 (2011) 0–3.

[29] C.E. Bannister, V.M. Lawson, Role of cement fluid loss in wellbore completion, Proc. - SPE Annu. Tech. Conf. Exhib. 1985-Septe (1985). doi: 10.2523/14433-ms.

[30] R. Bruckdorfer, A. Gleit, Static Fluid Loss Model, SPE Gen. (1988) 20. https://doi.org/.

[31] F. Sanfillippo, M. Brignoli, F.J. Santarelli, C. Bezzola, Characterization of Conductive Fractures While Drilling, SPE Eur. Form. Damage Conf. (1997) 319–328. doi:10.2118/38177-ms.

[32] R. Maglione, A. Marsala, Drilling mud losses: problem analysis, AGIP Internal Report, 1997.

[33] O. Liétard, T. Unwin, D.J. Guillot, M.H. Hodder, Fracture width logging while drilling and drilling mud/loss-circulation-material selection guidelines in naturally fractured reservoirs, SPE Drill. Complet. 17 (1999) 237–246.

[34] F. Civan, M. Rasmussen, Further discussion of fracture width logging while drilling and drilling mud/loss-circulation-material selection guidelines in naturally fractured reservoirs, SPE Drill. Complet. 17 (2002) 249–250.

[35] S. Sawaryn, Discussion of fracture width logging while drilling and drilling mud/loss-circulation-material selection guidelines in naturally fractured

reservoirs, SPE Drill. Complet. 4 (2002) 247–248.

[36] J. Huang, D.V. Griffiths, S.-W. Wong, Characterizing Natural-Fracture Permeability From Mud-Loss Data, SPE J. 16 (2011) 111–114. doi:10.2118/139592-PA.

[37] Majidi, S.Z. Miska, M. Yu, L.G. Thompson, J. Zhang, Quantitative Analysis of Mud Losses in Naturally Fractured Reservoirs: The Effect of Rheology, SPE Drill. Complet. December 2 (2010) 509–517. doi:10.2118/114130-PA.

[38] A.C. Hindmarsh, P.N. Brown, K.E. Grant, S.L. Lee, R. Serban, D.E. Shumaker, C.S. Woodward, SUNDIALS: Suite of nonlinear and differential/algebraic equation solvers, ACM Trans. Math. Softw. 31 (2005) 363–396. doi:10.1145/1089014.1089020.

[39] A. Nasiri, A. Ghaffarkhah, M. Keshavarz Moraveji, A. Gharbanian, M. Valizadeh, Experimental and field test analysis of different loss control materials for combating lost circulation in bentonite mud, J. Nat. Gas Sci. Eng. 44 (2017) 1–8. doi:10.1016/j.jngse.2017.04.004.

[40] S. Yousefirad, E. Azad, M. Dehvedar, P. Moarefvand, The effect of lost circulation materials on differential sticking probability: Experimental study of prehydrated bentonite muds and Lignosulfonate muds, J. Pet. Sci. Eng. 178 (2019) 736–750.

[41] M. Khafaqa, Experimental Study on Effectiveness of Lost Circulation Materials to Mitigate Fluid Losses, MONTANUNIVERSITÄT LEOBEN, 2016.

[42] R. Albattat, H. Hoteit, A Semi-Analytical Approach to Model Drilling Fluid Leakage Into Fractured Formation, 2020. https://doi.org/https://arxiv.org/abs/2011.04746.

[43] H. Hoteit, A. Firoozabadi, An efficient numerical model for incompressible two-phase flow in fractured media, Adv. Water Resour. 31 (2008) 891–905. doi:10.1016/j.advwatres.2008.02.004.

[44] B. Koohbor, M. Fahs, H. Hoteit, J. Doummar, A. Younes, B. Belfort, An advanced discrete fracture model for variably saturated flow in fractured porous media, Adv. Water Resour. 140 (2020) 103602. doi:10.1016/j.advwatres.2020.103602.

[45] M.B. Smith, C.T. Montgomery, Hydraulic fracturing, Crc Press, 2015. doi:10.1201/b16287.

[46] Z. Ben-Avraham, M. Lazar, Z. Garfunkel, M. Reshef, A. Ginzburg, Y. Rotstein, U. Frieslander, Y. Bartov, H. Shulman, Structural styles along the Dead Sea Fault, in: D.G. Roberts, A.W.B. T.-R.G. and T.P.P.M. Bally Cratonic Basins and Global Tectonic Maps (Eds.), Reg. Geol. Tectonics, Elsevier, Boston, 2012: pp. 616–633. doi:10.1016/B978-0-444-56357-6.00016-0.

[47] F. Irgens, Rheology and non-newtonian fluids, Springer International Publishing, 2014.

[48] D. Cioranescu, V. Girault, K.R. Rajagopal, Mechanics and Mathematics of Fluids of the Differential Type, Springer, 2016. doi:10.1007/978-3-319-39330-8.

[49] R.L. Panton, Incompressible flow, John Wiley & Sons, 2013. https://doi.org/https://doi.org/10.1002/9781118713075.ch10.

[50] T. Hemphil, A. Pilehvari, W. Campos, Yield-power law model more accurately predicts mud rheology, Oil Gas J. 91 (1993) 45–50.

[51] V.C. Kelessidis, R. Maglione, C. Tsamantaki, Y. Aspirtakis, Optimal determination of rheological parameters

for Herschel-Bulkley drilling fluids and impact on pressure drop, velocity profiles and penetration rates during drilling, J. Pet. Sci. Eng. 53 (2006) 203–224. https://doi.org/10.1016/j.petrol.2006.06.004.

[52] J. Lee, Well Testing, Society of Petroleum Engineers, 1982. https://store. spe.org/Well-Testing–P179.aspx (accessed June 29, 2020).

[53] R. Majidi, Modeling of Yield-power law drilling fluid losses in naturally fractured formations, The university of Tulsa, 2008.

[54] R. Majidi, S.Z. Miska, R. Ahmed, M. Yu, L.G. Thompson, Radial flow of yield-power-law fluids: Numerical analysis, experimental study and the application for drilling fluid losses in fractured formations, J. Pet. Sci. Eng. 70 (2010) 334–343. https://doi.org/10.1016/j.petrol.2009.12.005.

[55] R. Albattat, H. Hoteit, Modeling yield-power-law drilling fluid loss in fractured formation, J. Pet. Sci. Eng. 182 (2019) 106273. doi:10.1016/j.petrol.2019.106273.

[56] S. Baldino, S.Z. Miska, E.M. Ozbayoglu, A novel approach to borehole-breathing investigation in naturally fractured formations, SPE Drill. Complet. 34 (2019) 27–45.

[57] M.A. Ojinnaka, J.J. Beaman, Full-course drilling model for well monitoring and stochastic estimation of kick, J. Pet. Sci. Eng. 166 (2018) 33–43. https://doi.org/doi:10.1016/j.petrol.2018.03.012.

[58] Z. Yuan, D. Morrell, A.G. Mayans, Y.H. Adariani, M. Bogan, Differentiate Drilling Fluid Thermal Expansion, Wellbore Ballooning and Real Kick during Flow Check with an Innovative Combination of Transient Simulation and Pumps off Annular Pressure While Drilling, in: IADC/SPE Drill. Conf. Exhib., Society of Petroleum Engineers, 2016.

[59] Y. Sun, Near-Wellbore Processes in Naturally Fractured or Weakly Consolidated Formations Near-Wellbore Processes in Naturally Fractured or, (2017).

www.ingramcontent.com/pod-product-compliance
Lightning Source LLC
Chambersburg PA
CBHW081239190326
41458CB00016B/5848